SpringerBriefs in History of Science and Technology

More information about this series at http://www.springer.com/series/10085

Christine Yi Lai Luk

A History of Biophysics in Contemporary China

 Springer

Christine Yi Lai Luk
The Human and Social Dimensions
  of Science and Technology
  Ph.D. Program
Arizona State University
Tempe, AZ
USA

ISSN 2211-4564   ISSN 2211-4572   (electronic)
SpringerBriefs in History of Science and Technology
ISBN 978-3-319-18092-2   ISBN 978-3-319-18093-9   (eBook)
DOI 10.1007/978-3-319-18093-9

Library of Congress Control Number: 2015937741

Springer Cham Heidelberg New York Dordrecht London

Printed on acid-free paper

Springer International Publishing AG Switzerland is part of Springer Science+Business Media
(www.springer.com)

Luk's work problematizes "the West and the Rest" smugness about unitary trajectories for new (inter)disciplinary fields. Biophysics did not develop the same way in China as it did in the U.S. or Germany. The institutional constraints were not the same, nor was the government role in the development of the field analogous. Nevertheless, biophysicists managed to create a flourishing field that took account of and benefited from specific conditions in China.
—Ann Hibner Koblitz, Arizona State University

Too often, science policy and development policy assume that intellectual fields follow a single trajectory of development and therefore must be developed in new countries in the same way they've occurred elsewhere. As Luk's work on biophysics in China shows, however, this is simply not the case. Fields that ostensibly share the same name can take significantly different trajectories as they imagine and build themselves in diverse settings. Her work opens up the history of Chinese science & technology policy in

interesting and valuable ways, helping illuminate how China envisions and deploys scientific research as part of its social and political milieu.
　　　—Clark Miller, Consortium of Science, Policy, & Outcomes

This book is dedicated to Stuart M. Lindsay, who inspired me to write a book on biophysics

Acknowledgments

Many thanks are due to institutions and individuals for helping me along the way. I am grateful to have received the History of Science Society (HSS) and the National Aeronautics and Space Administration (NASA) Fellowship in the History of Space Science to pursue the research on which this book is based. The Dissertation Writing Fellowship awarded by the Graduate College of Arizona State University (ASU) gave me the time to revise my dissertation and turn it into a book manuscript. I appreciate the encouragement and support I have received from Ann Hibner Koblitz over the years. I wish to express my gratitude to Jane Maienschein, Clark Miller, and Hoyt Tillman for academic advice; to Rich Astone, Merrick Lex Berman, Monamie Bhadra, Shannon Conley, Lijing Jiang, Chad Manfred, Sharlissa Moore, Will Thomson, Brenda Trinidad, Daxue Wang for intellectual company; to Albert Spurgeon for reading and commenting on the manuscript; to Alexander Jost and Michael Wischnath for helping me to get access to some of the German materials; to Raffael Himmelsbach and Michael Markiw for helping me to comprehend and translate the German materials, and to my parents, for their care and love.

Finally, I would like to thank Lucy Fleet, Cynthia Kroonen, and Ayswaraya Nagarajan for their editorial assistance and an anonymous reviewer for making publication with Springer possible. I take full responsibility for any mistakes that remain in the book.

Contents

Prologue

In the late summer of 2011, I found myself at a faculty–graduate student potluck party. As I was walking down the hallway trying to make conversation, I stumbled across a middle-aged faculty member who taught general education courses at my university. After I had told him of my interest in doing a dissertation on the history of biophysics, he mentioned to me that he held a doctorate in biophysics from the University of Virginia. He asked about my motives for studying biophysics because, unlike him, I did not have a degree in that field nor was I pursuing one.

"I am intrigued because it seems that no two biophysicists can agree on what biophysics is specifically, yet it is held that biophysics is important and worth doing." To which he replied in jest, "True. We don't know what biophysics is exactly, but that's what we do."

Contemporary history of biophysics appears to be in a perpetual state of perplexity. In 1940, J.R. Loofbourow from MIT suggested that there is "no clear agreement, even among biophysicists, as to what the term biophysics means." (Loofbourow 1940, p. 267). He then took the liberty to expand the conceptual scope of biophysics to include the physics of life, the physical methods to study biological systems, and the physical intervention in life processes. What Loofbourow illustrated in 1940 was the broad scope of the field while lamenting the lack of a definable focus of biophysics.

The molecular revolution since the late 1940s and early 1950s changed little about this picture. The discovery of DNA and the advancement of microscopic instruments did indeed improve the tricks of the trade but these scientific developments still did not alter the fundamental set of questions regarding the conceptual problem of the focus of biophysics. By 1963, a *Nature* article kicked off with a report on the organization of biophysics in the international realm with the following lines, "probably no two scientists would agree on a definition of 'biophysics', yet during the past five or ten years there has been a growing consciousness that in this ill-defined borderland where physics, physical chemistry, biology and medicine overlap, revolutionary advances are likely to be made in the next ten or twenty years." (Sutherland 1963, p. 1141). In 1969, *Nature* continued to portray biophysics as an amorphous but attractive discipline (*Nature* 1969).

Fast-forwarded to the 1980s, one would expect the definitional problem of biophysics to be a bygone issue but that was not the case. One unanimous conclusion arrived at in the Eighth International Congress of Biophysics held in Bristol between 29 July and 4 August 1984 was that "there probably never was an adequate definition of biophysics. Perhaps there never will be." (Clarke 1984, p. 605). The same year, a book reviewer commenced a review article of a biophysics textbook with the following sets of questions:

> What is biophysics? This question, asked by the editors of *Biophysics*, must also have been asked countless times of all of us who profess to be biophysicists. To develop a sense of unity and to delineate boundaries is perhaps the main obstacle to be surmounted by anyone who seeks to write a core textbook on this discipline. Yet, the writing of such a text would in itself represent an essential stage in the definition of the subject. Biophysicists have long awaited such a statement—have the authors of *Biophysics* provided it? (North 1984, p. 755).

The reviewer concluded that the authors of the textbook under review gave a satisfactory attempt at defining the scope of biophysics but not a "definitive statement that we have been awaiting." (North 1984, p. 756).

In 1996, comparing definitions and coverage of biophysics in different textbooks, Marco Bischof from the International Institute of Biophysics in Germany concluded that he could not find a common definition of the investigative range of biophysics in two biophysics textbooks (Bischof 1996).

In other words, from the 1940s to the 1990s, no one had been able to offer a textual definition to adequately cover biophysics in spite of the proliferation of textbooks and handbooks dedicated to the subject itself. Instead, one article on *Nature* suggested that the question of the definition of biophysics should be approached by seeking what biophysicists do in practice (Miller 1985).

The motto "definition by doing" harkens back to the same sentiment echoed by the self-identified biophysics faculty member when he told me that not knowing what biophysics is does not prevent a biophysicist from doing biophysics.

This is a study about the history of biophysics—a subject that has fascinated me ever since the first semester of my doctoral program. I had no previous training in the science of biophysics, nor did I have an obsession involving the latest microscope technology, nor did I have any familial, cultural, or personal predilection for biophysics. What fascinated me was that biophysics did not have a definite set of subfields or a university-wide department of its own in the United States. Different academic institutions invented their own rules to incorporate biophysics into different schools and departments, and different people used terms like "biophysics" and "biophysicists" to mean different things. Yet, despite the lack of a formal definition in this field, there was nevertheless a great deal of optimism surrounding its future.

References

Bischof M (1996) Some remarks on the history of biophysics (and its future). In Zhang CL, Popp FA, and Bischof M (eds). Current development of biophysics: the stage from an ugly duckling to a beautiful swan. Hangzhou University Press, Hangzhou, pp 10–21

Clarke M (1984) Many guises in bristol. Nature 31(15987):605

Loofbourow J (1940) Borderland problems in biology and physics. Rev Mod Phys 12:267–358

Miller A (1985) Definition by doing. Nature 317(6035):300

Nature (1969) Interdisciplinarians: bright future for biophysics. Nature 223 (5213):1317

North ACT (1984) In search of a subject: book review of Biophysics edited by Walter Hoppe, wolfgang lohmann, Hubert Markl and Hubert Ziegler. Nature 308(5961):755–756

Sutherland GBBM (1963) International organization of biophysics. Nature 198(4886):1141–1142

About the Author

Christine Yi Lai Luk holds a doctorate in the human and social dimensions of science and technology from Arizona State University (ASU). She is interested in exploring the intersection between the scientific establishment and cultural knowledge in non-Western settings. Her pursuit of non-Western scientific culture was supported by the History of Science Society (HSS) and the National Aeronautics and Space Administration (NASA), the D. Kim Foundation for the History of Science and Technology in East Asia, and the Sir Edward Youde Memorial Fund Council. This book stems from her doctoral dissertation by the title of "Biophysics, Rockets, and the State: the Making of a Scientific Discipline in Twentieth-Century China."

Before coming to the United States, Dr. Luk completed her M.Phil. at the City University of Hong Kong where she taught a variety of courses ranging from globalization and development, social change and development in East and Southeast Asia. Her work has appeared in *Journal of the History of Biology*, *Research Yearbook of Philosophy of Technology in China*, and *Women in Engineering and Technology Research*. Her research endeavors are centered around the re-examination of historical relationships and the reimagination of cultural patterns through the prism of biomedical science in East and Southeast Asia.

Chapter 1
Introduction: Biophysics in Contemporary China

Abstract Following the prologue that introduces amorphousness and fluidity as central characteristics of biophysics in Western scientific literature, this chapter argues that since biophysics has always been known as a discipline "defined by doing," it is meaningful to study the varying content and the formational context of the discipline. The foundation of my study is: as biophysics can only be captured through depicting what biophysicists do in practice, the Chinese experience of building biophysics is one of the many ways that gives a concrete meaning to biophysics. Biophysics was in its infancy in the first half of twentieth-century China, but its connection with the military-industrial complex gave it the necessary financial resources and political impetus to survive and thrive. In addition to a chapter-by-chapter summary, this chapter considers the peculiarities of China's biophysics as it was immersed in the experimental development of research rockets for the strategic defense program, *liangdan yixing* (*Two Bombs, One Star*).

Keywords Liangdan yixing · Definitions of biophysics · Disciplinary characteristics · Nature · Origin of China's human spaceflight

On 2 November 2009, Yang Liwei (杨利伟), China's first man-in-space, and Chen Shanguang (陈善广), director of the China Astronaut Research and Training Center and chief engineer for the Chinese manned space program, dressed in military uniforms to attend a funeral of a deceased scientist. The scientist to whom these two space celebrities paid respect was not a Werner von Braun type of figure in China, and not a typical space pioneer you'd expect by any measure, because he was a cell biologist (IBP-CAS 2009).

The reasons driving the first *taikonaut* and the director for training *taikonauts* to pay homage to a biologist are actually quite simple. Forty-three years ago, Bei Shizhang (贝时璋), in his capacity as the founding director of the Institute of Biophysics at the Chinese Academy of Sciences, was responsible for launching the first biological sounding rockets carrying animals to the upper atmosphere. Although the Chinese space travel in the sixties only brought dogs and mice to reach a level not even close to low earth orbit, many Chinese regarded the endeavor

© The Author(s) 2015
C.Y.L. Luk, *A History of Biophysics in Contemporary China*,
SpringerBriefs in History of Science and Technology,
DOI 10.1007/978-3-319-18093-9_1

as the origin of human spaceflight in modern China.[1] Therefore, the high-profile visit to Bei's funeral by lieutenant Yang and administrator Chen was a symbolic act to acknowledge the role of Bei Shizhang and the Institute of Biophysics under his directorship in setting the precedent for China's human spaceflights.

The history of biophysics in China seems to be a far cry from the debates about the definition of biophysics which I discussed in the prologue. But if biophysics can only be captured through depicting what biophysicists do in practice, the Chinese experience of building biophysics by launching rockets is one of the many ways that gives a concrete meaning to biophysics. How has the practice of Chinese biophysicists shaped the definition of biophysics in China? What forms of knowledge are involved in articulating the subject matter of biophysics? For those who are relatively unfamiliar with Chinese history, this study will offer an opportunity to explore contentious issues regarding the history of a scientific discipline in a different socio-historical context. China scholars and Chinese readers, I hope, will find that Chinese history looks quite different when the issue of building a scientific discipline is placed at the center of the analysis.

In the history of science, it was Robert Kohler (1982) who advocated approaching the building of scientific disciplines from a socio-economical point of view. His compelling analysis of the historical growth of biochemistry in the US indicates that the social and institutional contexts matter to the expansion of a scientific discipline as much as, if not more than, the intellectual contributions and conceptual underpinnings of a discipline.

As Kohler carefully justified the role of institution in shaping science, a bigger question remains as to whether my analysis of the Chinese history of biophysics adds anything to the scholarship. If the building of a scientific discipline depends on institutional backing, is the character of institutional shaping a universal attribute or does it exhibit distinct, tangible national difference?

Jane Maienschein (1985) has speculated whether national setting might exert any discernable effect on the development of scientific activities. As she explained, although institutional studies that focused on personalities or major historic events have left out the intellectual or political origins of the institutions, they still offer considerable insights into what role the national setting has played in shaping the character of science. Thus it is possible and sensible to argue for a certain "self-consciously American biology" in its own right.

Is there a distinctively Chinese biophysics? At various points, I compare and contrast the Chinese efforts of building biophysics with the American historical counterparts in order to put the Chinese experience of instituting biophysics into perspective. Using postwar American history of biophysics as a control case, the comparative analysis shows that there are indeed notable differences in constructing biophysics in twentieth-century America and China.

---

[1]Wang Xiji (王希季), former director of the China Academy of Space Technology (CAST), applauded the efforts of the biophysicists this way: "the research of cosmobiology and the launch of biological rocket flights at the Institute of Biophysics laid the cornerstone for developing China's manned spaceflight." See IBP-CAS (2009) foreword.

Structural characteristics of disciplinary orientation and university opportunities shape the history of biophysics in twentieth-century China, but they played out in distinctive ways in the context of China apart from that of East Asia. Disciplinary orientation, and with it ways of ordering scientific knowledge and practice, came in mostly with the national mobilization of personnel and capitals. Science stood for social progress and public goods. Opportunities to promote organized activities of science, in terms of creating research institutes and educational programs, became more available after the founding of a unified state in 1949.

There are several central themes in the Chinese narrative of biophysics that are unfamiliar to many Western readers. In China, biophysics thrived as part of the space program. This is illustrated in the interesting scene of the 2009 funeral of Bei Shizhang in which China's first man in space and the director of China's manned space program were honorable guests. The particular setting was to highlight the importance of the socio-historical context for the contents of biophysics in contemporary China. It was under the specific historical circumstances where Chinese biophysicists pioneered human spaceflight through putting animals to the sky that allowed the combination of biophysics and space exploration.

Nevertheless, Chinese biophysicists could not have succeeded in launching the biological sounding rockets without the political patronage of the state and cooperation with other actors in the space-and-missile program, with one striking feature being the state's direct, sustained involvement in building the space establishment through mobilizing experts and laymen from many different areas.

There are some similarities between the Chinese space operations and those in the West. In both, the role of the state is critical, especially in the beginnings of space programs, as space science is inextricably intertwined with the development of missiles, nuclear weapons, and issues of national security in general. Oftentimes, the state plays an elemental role in unifying the various forces in the military and civilian sectors for doing rocket research. While similar in that respect to its Western counterparts, space operations in China have been more centrally organized with far-reaching political ramifications.

In China, spacecraft and statecraft are intertwined to the extent that technical advancement in spacecraft reflects the governing capabilities of the state. The state participation in China's rocket-and-space program led Qian Xuesen, considered the "father of missiles and rockets in China," to the belief that "our system can effectively integrate and unify our willpower. This is more conducive for conducting rocket engineering than in liberal America." (Gong 2006, p. 26). Qian was making not a scientific observation but a political statement on the credibility of the socialist system and the competence of the party-state in leading the country to extend the frontiers of science.

The explicit attention given to the political implications of space development is remarkable but hardly astonishing to Western scholars, being reminiscent of what Yaron Ezrahi has observed about the political appropriation of science and technology in both liberal and non-liberal political traditions. Regarding the function of science in non-democratic contexts, he wrote:

The very science and technology which authorize decentralization by specialization, because they substantiate instrumental rather than arbitrary or political grounds for unifying parts of action, can also authorize centralization. In a totalitarian state where no autonomous private sector exists, the employment of science and technology to legitimate centralized political control in terms of necessary technical unity is not mitigated by the ideologically sanctioned decentralizing effects of specialization, the authority of nonpublic bodies, and the public nature of science as an intellectual enterprise (Ezrahi 1990, p. 44).

Ezrahi was talking about the politicization of science not in China per se but in countries he considered to be "totalitarian states" such as the Soviet Union, Nazi Germany, Franco's Spain, and fascist Italy. On the topic of seeking political legitimation through borrowing the authority of science and technology, H. Lyman Miller has considered the political manipulation of science and technology in modern China:

One way in which science has had a significant political impact in modern China has been in its appropriation by intellectuals as a basis for comprehensive political ideologies. Because science has had spectacular success in explaining aspects of the natural world, it has laid claim to a uniquely reliable kind of knowledge and become a powerful source of ideas and values. Thus, science has acquired a sometimes intimidating prestige and authority, leading at times to attempts to extend its methods beyond its domain (Miller 1996, pp. 4–5).

Questions as to whether People's Republic of China (PRC) can be considered as a "totalitarian state" or the extent to which science justifies politics in PRC are beyond the scope of this study. It is not my purpose to criticize either Ezrahi or Miller with these short quotations, but it is my intention to highlight some of the prevailing scholarly understandings of the function and value of science in an authoritarian/totalitarian non-Western country like China. While I do agree that science and technology sometimes serve as tools at the hands of the governing bodies to accomplish military and political goals, ideological legitimation is not sufficient to explain the gamut of activities of science and technology in China. To write off the efforts of Chinese scientists as constituting nothing more than a co-optation with the state agenda to manipulate the public is to ignore many other important aspects in the science and technology sector.

The primary actor, sponsor, and orchestrator of the Chinese space program is undeniably the state and its many apparatuses. But the space program has also enjoyed substantial popular support from its early days, a trend which continues into the present. Nationalist sentiment (in various forms and brands) is a necessary factor here, though insufficient by itself since "it would be difficult to imagine nationalism succeeding without inspiring personal attachment" as Sigrid Schmalzer sharply noted in her study of the popular roots of human fossils and human evolution in modern China (Schmalzer 2008, p. 278). Likewise, it is problematic to reduce the significance of the Chinese space program to factors like nationalism or cultural chauvinism alone. Starting from the very early days, Chinese rockets have been dedicated to the needs of the masses on a practical level. The *Fengyun* meteorological satellites fulfilled cultural responsibilities of weather monitoring and disaster management. Premier Zhou particularly emphasized the development of indigenous satellites and utilization of foreign satellite data; Deng Xiaoping encouraged the marketization of communication satellites. Stacey Solomone (2012, 2013) has recently identified the linkage that the Chinese government has fostered between the aerospace industry and popular demands:

> in order to serve the people, space programs were vectored toward communications, entertainment, and life science applications. These applications of the aerospace industry into the lives of the masses are how the Chinese maintain a direct connection to space technologies; this approach reaches beyond the goals of national defense and international prestige...Although the original goals of national defense and international prestige remain strong, the aerospace industry of China also seeks to serve the people and contribute to China's economic development (Solomone 2012, p. 245).

In addition to "serving the people" through providing forecasting and infotainment services, "serving the science" is also an implicit outcome of the space program. I argue that advancement of scientific disciplines is one of the oft-ignored contributions of the space-and-rocket missions. The space program created a political platform to facilitate the growth of space-related disciplines in China— biophysics being a case in point. From this awareness of the interplay between biophysics, rockets, and the nature of the state of contemporary China, we can move to the next focus, the father of biophysics in China.

1.1 The Father of Biophysics in China

In the prologue, I suggested that Western biophysicists have not arrived at a unanimous agreement as to what biophysics really is; rather, it is suggested that doing biophysics is more important than clearly defining it. That practice precedes definition might work for some practicing biophysicists; but a conceptual foundation is indispensable for a discipline to grow and develop as a profession. Scientists defining their disciplines rest upon their scientific credibility and prestige.

Bei Shizhang (1903–2009) was the most authoritative spokesperson of biophysics in China. He was the most influential biophysicist who has voiced opinions about what biophysics encompassed in China. As the most respected leader in the community of Chinese biophysicists, he was considered to have almost single-handedly founded biophysics in the PRC. Several questions emerge regarding this towering figure in coordinating the disciplinary formation of biophysics. What is his academic background and area(s) of expertise? How does his professional background correlate with his decision-making capacity and authority as a discipline leader? Given the knowledge of Bei's perception and predilection, what political and economic resources were available to him to achieve his discipline-building project?

Answers to these questions are attempted in Chap. 2. Central to understanding the historical trajectory of biophysics in contemporary China is the intersection between individual scientists and macro-historical events. The life of Bei Shizhang and his discipline-building dream operate in a context of political expectations, economic constraints, and competing social and scientific projects. These structural features condition the perceptions and aspirations of Bei and determine the disciplinary pattern. Chapter 2 elucidates the connection between Bei's professional life, sources of influence, controversy over his scientific achievement. At the heart of the analysis is his lifelong struggle to put forth an objectionable theory of cell reformation.

1.2 The Institutional Infrastructure of Biophysics

There are many methodological strategies to examine the construction of a scientific discipline. Kohler's pioneering study of the ascendancy of biochemistry considered the emergence of departments and institutions, the founding and appointment of chairs, specialized journals and societies, the standardization of degrees and certificates, and the creation of academic markets for university graduates as indicators of the consolidation of biochemistry in the US compared to Europe. What Kohler highlighted was the significance of institutional opportunities in shaping the medical programs in comparative Euro-American contexts (Kohler 1982).

In his study of the medical policy and institution in the PRC, David Lampton proposed the following variables be included in his theoretical framework: construction of medical institutions, medical education, distribution of services, types of research, and guidelines for professional development (Lampton 1977). Although issues of policy-making in medical and health service are not the focus of my study, Lampton's methodological advice is helpful for identifying the salient characteristics of the scientific and biomedical institution in post-1949 China.

My inquiry of the disciplinary formation of biophysics combines the methodological perspectives offered by Kohler and Lampton. Since institutional contexts provide material and political support for certain scientific disciplines to grow, I attribute a considerable part of the constitution of biophysics to its institutional infrastructure. In my study, I examine the institutional infrastructure in terms of the construction of research institution, educational system, and types of biophysical research and services, as well as activities for furthering the professional life of biophysicists. Chapter 3 looks at how biophysics coalesced into a distinct discipline through the institutionalization of teaching, research, journalistic, and professional development of biophysics. I argue that the history of instituting biophysics in contemporary China can be assessed from the establishment of educational and research programs to the specialized journal and professional society for biophysics. For readers interested in the institutional history of science in non-Western countries, this chapter will attempt to shed light on how a science discipline takes root in different institutional contexts. Conversely, historians of education in modern China can determine the similarities and differences between the teaching of scientific and non-scientific disciplines in twentieth-century China.

1.3 Building Biophysics Through Launching Sounding Rockets

Western space observers have not normally grasped the connection between Chinese biophysicists and the Chinese space program. Although the first manned orbital flight in 2003 aroused much attention, there is a considerable conundrum as to what came before. Reviewers of space science recognized the launch of China's first communication satellite (comsat) in 1984, or the launch of the country's first satellite, *The East is Red*, in 1970, and perhaps the tale of Qian Xuesen who returned to China in 1955 (Kulacki and Lewis 2009). But a closer inspection reveals significant analytical gaps as a unique *biological* dimension was missing among the known space vehicles and dignitaries. The comsat has no living things in it and only feeds electronic data back to earth; the *East is Red* is bigger and heavier than the Sputnik, but it is nothing more than an empty vessel circling the earth making noise; Qian is undeniably a pivotal figure in China's aerospace industry but he is not primarily responsible for putting any forms of life into space as his expertise is not in biology. Although Qian was credited for founding the missile-and-space program in China and having written the popular textbook *An Introduction to Interplanetary Flight* (星际航行概论), the root of a manned spaceflight can be traced back to the large-scale bomb-and-rocket mission dubbed "Two Bombs, One Star" (两弹一星, referring to the detonation of an atomic bomb, a hydrogen bomb, and the launch of an artificial satellite).

A conventional view of "Two Bombs, One Star" is that political missions precede disciplinary interests. This aspect of the relationship between discipline and mission was captured in the slogan "mission drives discipline" (任务带学科). But the complementary aspect was usually overlooked, for "mission drives discipline" was just the upper line in the slogan, to be completed by the lower line——"discipline facilitates mission" (学科促任务).

Existing writings on the history and technology of "Two Bombs, One Star" have been mostly provided by retired rocket engineers, nuclear scientists and government officials. The focus is either on the technical capability or the political correctness of the program. Few looked at the program from the perspective of discipline building. Space missions came to fruition with a close interaction and cooperation among physical scientists, engineers, life scientists, mission planners, cadre members, officials, and technicians. Since mission planners and government officials dominated the narratives of the event, we have yet to understand and appreciate the role of biological scientists in the mission or what the mission meant to the biology-related disciplines.

In light of this conceptual gap, Chap. 4 is dedicated to elucidate ways in which "Two Bombs, One Star" generated the political momentum to integrate biophysical scientists into the overall mission and, in the process, created the operative condition for the active participation of life scientists in the space program, most notably in the launch of biological sounding rockets in the 1960s. Essentially, this is an examination of how the contribution of biophysicists to "Two Bombs, One Star" in the 1960s shaped the discourse and character of biophysics in China after the completion of the mission.

1.4 Overview

The narrative that follows will cover the history of biophysics in China roughly from 1930 to 1980. Each of the ensuing chapters investigates the core components I identified as the focus of my research. Chapter 2 gives a biographical sketch of Bei Shizhang, the founding father of biophysics in modern China, with a chronicle of his academic background with his scientific style and worldviews; Chap. 3 explores the institutional infrastructure of biophysics in areas of research, teaching, technical journal, professional society; Chap. 4 takes up the issue of the political patronage of biophysics by tracing the military-scientific negotiations in the early periods of China's rocket research; finally, Chap. 5 summarizes the intellectual landscape of biophysics in contemporary China. I also identify the research significance and qualification through identifying the disciplinary strengths and weaknesses of biophysics in contemporary China.

References

Ezrahi Y (1990) The descent of icarus: science and the transformation of contemporary democracy. Harvard University Press, Cambridge

Gong XH (巩小华) (2006) Inside the decision-making world of Chinese space industry (中国航天决策内幕). Chinese Literature and History Press, Beijing

IBP-CAS (2008) Flying dogs in the sky: a documentation of Chinese biological experimental rockets (小狗飞天记: 中国生物火箭试验纪实). Science Press, Beijing

IBP-CAS (2009) Representatives of the China astronaut research and training center Chen Shanguang and Yang Liwei send condolences to academician Bei Shizhang's family (中国航天员中心陈善广、杨利伟一行吊唁贝时璋院士并慰问家属). http://159.226.118.206/detailt.aspx?newsno=10323. Accessed 15 Dec 2013

Kohler R (1982) From medical chemistry to biochemistry: the making of a biomedical discipline. Cambridge University Press, Cambridge

Kulacki G, Lewis JG (2009) A place for one's mat: China's space program, 1956–2003. American Academy of Arts and Sciences, Cambridge

Lampton D (1977) The politics of medicine in china: the policy process, 1949–1977. Westview Press, Boulder

Maienschein J (1985) History of biology. Osiris 1(2):147–162

Miller HL (1996) Science and dissent in post-mao China: the politics of knowledge. University of Washington Press, Seattle

Schmalzer S (2008) People's Peking Man: popular science and human identity in twentieth-century China. University of Chicago Press, Chicago

Solomone S (2012) Space for the people: China's aerospace industry and the cultural revolution. In Brock D, Wei CN (eds) Mr. Science and Chairman Mao's cultural revolution: science and technology in modern China. Lexington Press, Lanham, pp 233–250

Solomone S (2013) China's strategy in space. Springer

Chapter 2
The Father of Biophysics in China

Abstract This chapter intends to elucidate the academic heritage and intellectual background of the father of biophysics in China. My central argument is that if we want to learn about biophysics in China, we have to learn more about Bei Shizhang; if we want to understand Bei, we have to pay a visit to his intellectual world. But this world is complex and many-faceted, for Bei is not just an institution builder and a visionary leader, he is also a charismatic leader and a controversial scientist. This chapter explores these various facets of Bei Shizhang by chronicling his early life in China, the correlation between the formation of his scientific worldview and the *naturphilosophie* of his neo-Lamarckian German teacher, Wilhelm Harms. I also consider Bei's contested scientific pursuit—the theory of cell reformation which had some parallels with O.B. Lepeshinskaya's now discredited studies of "the origins of life" in the Soviet Union—within the larger sets of issues involving Soviet-styled biology, the perception of Marxist philosophy, and political patronage of cytology in twentieth-century Communist states.

Keywords Bei Shizhang · Cell reformation · Father of biophysics in China · Tübingen · Wilhelm Harms

When Bei Shizhang passed away in Beijing in 2009, there were two major details that captured the media's attention.[1] First, at the age of 106, he was the oldest academician in modern China, being among the first batch of scientists to be elected to the *Academia Sinica* in 1948, right before the Nationalist government retreated to Taiwan. He was the last academician whose scientific membership straddled across the Nationalist and Communist regimes, and after 2004, Bei became the only surviving academician who witnessed the transformation of modern China from the "Republic of China" to the "People's Republic of China." The second detail was that he was lauded as the founding father of biophysics in China. People were told that biophysics could not have materialized in China without him. Biophysics and

---

[1]There were many news reports, videos, and articles between October and November 2009 on Bei's death and legacy. See the following for some of the most representative ones: Committee of the Funeral Service of Bei Shizhang (2009), Zhao and Chen (2009), *China News* (2009), *Xinhua News* (2009), *Science Times* (2009), *People's Daily* (2009).

© The Author(s) 2015
C.Y.L. Luk, *A History of Biophysics in Contemporary China*,
SpringerBriefs in History of Science and Technology,
DOI 10.1007/978-3-319-18093-9_2

Bei Shizhang were synonymous in China, and this was represented through Bei's pioneering efforts in synthesizing physical and biological sciences (Wang 2009; *Science Times* 2010).

In 2003, a *festschrift* commemorating his centennial birthday entitled *Bei Shizhang and Biophysics* was released. Several of his students and colleagues enumerated his tremendous contribution to laying the groundwork of biophysics in China. The biographical narrative did not merely center around the personal virtues of Bei as an indefatigable scientist, but also as a deep thinker, an institution builder, and a visionary disciplinary leader. These various facets of his life painted a human picture of Chinese biophysics, as Bei's intellectual commitment and political conviction personified the biophysics program in China.

I assume most English readers have never heard of Bei Shizhang, let alone his contribution to biophysics. Therefore, this chapter will give a general biographical sketch of Bei Shizhang stressing his intellectual commitment and academic heritage. It is my assertion that if we want to learn about biophysics in China, we have to learn more about Bei Shizhang. In what follows, I focus on the academic background and scientific contributions of Bei Shizhang—the first and arguably the most important biophysicist in China—in order to investigate the epistemic origins of biophysics in contemporary China.

2.1 Bei Shizhang: the Early Years

The early life of Bei Shizhang, as chronicled by Bei's student Ying (1992) spanned from his birth in 1903 to his return to China from Tübingen in 1929.[2]

The first stage of Bei's life stretched from his childhood to his undergraduate education in China and graduate training in Germany, which was chronicled by Ying Youmei, while Zhang Jianjun (张坚军) presented a much simplified version of Bei's pre-schoolings and familial life intended for juvenile readers (Zhang 2002). Both accounts were written in Chinese. For the purpose of my analysis, the most important round of events in this stage was his graduate experience in Germany. Below I will reiterate only important milestones in his pre-German phase.

In 1903, Bei Shizhang was born to a fisherman's family in the coastal village named Zhenhai (镇海) in Zhejiang province. As the name of the fishing village implies—Zhenhai means literally to conquer (*zhen*) the sea (*hai*)—it lies along the seaside frontier facing the South China Sea, overlooking the Formosa Strait. Bei came into this world at an unpropitious time, for China was steeped in the

---

[2]In chronicling Bei's life and work, Bei's student Ying Youmei divided his timeline into three stages: the first stage spanned from his pre-German schooling and German education to his return to China where he was appointed as chair of the biology department at Zhejiang University (*Zheda*), a position he held from 1930 to 1949. The *Zheda* period was stage two. In 1950, he moved to Beijing to establish and organize the Institute of Biophysics of the Chinese Academy of Sciences. That was stage three.

interregnum period after the Boxer Uprising and before the establishment of the Republic of China. Bei was the first one in his family to get formal schooling. As much as he enjoyed eating salted fish and toying with herring, he knew from an early age that getting good grades were far more important than knitting a tight seine. He had witnessed how hard his parents toiled day and night, on and off the boat, making barely enough money to make ends meet. Life staged on the maritime and wartime theatres was harsh and demanding.

Although he had to swallow more pain than food at an early age, he had always believed in the promising engine of education fueled by science. The primary and secondary schools he had attended, namely the Jin Xiu School (进修学堂), Bao Shan School (宝善学堂), and De Hua School (德华学校) were German-run charities in China, so he had studied the German language for several years before going off to college. His father encouraged him to further study foreign language, but Bei was certain that his real interest was in life science. In 1919 he was admitted to the Tongji Technical College of Medicine and Engineering, TTCME (同济医工专门学校), formerly the German-founded Tongji School of Medicine and Engineering (同济医工学堂) and graduated with a pre-medical degree in 1921.

His college completion coincided with the defeat of Germany in World War I. The currency depreciation after the war made graduate education in Germany affordable for Bei. It was estimated that his tuition fee in post-WWI Germany was almost the same as that of TTCME. Along with his two friends from TTCME, he boarded the ocean-liner *Amazon* bound for Marseilles in the summer of 1921. Their travels into inland France finally landed them on a train bound for Freiburg (Wang 2010a, pp. 19–25).

The three Chinese students chose the Universität of Freiburg because their pre-medical credentials from TTCME were accepted by Freiburg. Bei's TTCME companions Yu Hongkang (余鸿康) and Li Yuanshan (李元善) applied immediately to the School of Medicine at Freiburg but Bei applied to the School of Natural Philosophy instead. Although he held a pre-medical degree, he had always been interested in a broad-based education in natural science. He took advantage of this opportunity to take various classes by renowned scholars in their fields, including zoology with Nobel-prize winner Hans Spemann, botany with Friedrich Oltmann, physics with Franz Himstedt, chemistry with the Nobel laureate Heinrich Otto Wieland, anatomy with Eugen Fischer, and pathology with Ludwig Aschoff.

In 1922, he moved to Münich hoping to study with the famous zoologist Richard Hertwig. He continued his multifarious coursework at the Universität der München, taking chemistry with Nobel-prize winner Richard Willstätter, physics with Wilhelm Wien, botany with Karl von Göbel, paleontology with Ferdinand Broili, and geology with Emmanuel Kayser. Years later, his multi-disciplinary transcript would prove useful for building the Institute of Biophysics in China as the next chapter will disclose.

Bei's original plan was to undertake a doctoral project under the supervision of Hertwig. As Hertwig was about to retire, he referred Bei to Friedrich Blochmann at Tübingen. So, in 1923, Bei left the cosmopolitan city of Münich for the small university town of Tübingen. Yet it turned out that Blochmann was going to retire too. Bei was directed to conduct experiments on the cell constants (*zellconstanz*) of the parasites *oxyuris obvelata* and *o. tetraptera* under the supervision of Richard Vogel for two years until the arrival of Wilhelm Jürgen Heinrich Harms.

2.2 Wilhelm Harms

It was a long journey before Bei could embark on his PhD research. From Freiburg to München to Tübingen, from Spemann to Hertwig to Vogel, he finally settled down with Wilhelm Harms as his advisor. According to Ying's documentation, Harms was one of the earliest surgeons in sex reassignment operations. Along with Voronoff and Steinach, Harms was famous for their pioneering work on "rejuvenation" research. Ying drew upon his writings on various methods of "rejuvenation" from his two-volume *Body & Germ Cells* (*Körper und Keimzellen*) published in 1926. In the first volume, Harms reviewed the nature and history of germ cells and their interconnections with the body.

A germ layer is a group of germ cells formed during embryogenesis. The germ layer theory was the cornerstone of Ernst Haeckel's biogenetic laws, in which he maintained that only after the germ layers were formed in the gastrula was there structural differentiation. The biogenetic theory was controversial partly because it dealt with the relationship between individuals and groups (or ontogeny and phylogeny). Haeckel saw individual organisms repeating the ancestral reminiscence in the course of development, which was known as "recapitulation theory." The theory was chiefly about to what extent features were inherited from our ancestors and which were epigenetic adaptations to external environment.

Maienschein (1978) has assessed the opposing views expressed by Wilhelm His, E.B. Wilson, E.G. Conklin, and F.R. Lillie as to whether "ontogeny recapitulates phylogeny." According to Maienschein, Wilhelm His subscribed to hereditarian thinking in explaining embryonic development, while E.B. Wilson held a more balanced view of the relative roles played by hereditary and epigenetic factors. Both His and Wilson denied a "true recapitulation" of ontogeny by phylogeny.

More relevant to Harms' proposition are E.G. Conklin and F.R. Lillie. E.G. Conklin shared Harms' interests in the origin and formation of the germinal layers. Like his contemporaries, Conklin (1897) was drawn to the developmental history of fertilized eggs. But his investigative lens was placed under the cytological changes that occur during cellular differentiation. The significance Conklin attached to cleavages stemmed from his assertion that a comparative inquiry of forms and causes of cleavages in closely related organisms would elucidate the homologies of blastomeres in different animals. Conklin distinguished between different types of cleaving ova among various types of organisms. He considered the various types of cleavages (determinate and indeterminate cleavages) as a reflection of diversity in the natural world, for "Nature is continually performing some very remarkable experiments in her own way." (Conklin 1897, p. 4). Conklin found E.B. Wilson's sweeping generalization of the "hereditary tendency" to all organisms questionable.

Conklin showed that a comparative study of cell lineages in closely and not-so-closely related organisms was valuable to clarify the ontogeny-phylogeny relations. The last one in Maienschein's cell-lineage circle was F.R. Lillie. Lillie also objected to E.B. Wilson's omission of cleavage heterogeneity at the early dividing stages. Drawing from his studies on the life history of the mussel *Unionidae*, Lillie argued that the early cleavage stages were not simply ancestral reminiscences but adaptive features to the later adult stages. Both Conklin and Lillie insisted that the early differences in cleavage were critical to later development. Maienschein carefully showed that despite their minor disagreements, these four biologists found Haeckel's biogenetic laws inadequate to explain the ontogeny-phylogeny relations. Ontogeny did not just "recapitulate" phylogeny. As Lillie summarized, "The ontogeny is inherited no less than the adult characteristics, and is subject to precisely the same laws of modification and variation." (quoted in Maienschein 1978, p. 153)

Against this historical backdrop of ontogeny-phylogeny debate, we can now proceed to consider some of Harms' legacies. Compared to the American orthodoxy on ontogeny, Harms' takes on the phylogenetic issues were somewhat heretical. To begin with, Harms was an outspoken "neo-Lamarckian" when it came to *"des Erwerbs neuer adaptiver Eigenschaften"* (the acquisition of new adaptive features). Hull (1984) has pointed out that it was inadequate, sometimes misleading and even defamatory, to call someone a "Lamarckian" not only because "nearly every type of hereditary phenomenon has been termed at one time or another Lamarckian," but also because "at times Lamarckian was a pejorative term to be used to characterize the views of one's opponents." (Hull 1984, p. xliii). In the case of Harms though, the label "neo-Lamarckian" was given to Harms not by me but by others (Potthast and Hoßfeld 2010). But I agree with Hull that what was more meaningful than borrowing the name of Lamarck was an elaboration of the specific "Lamarckian" or "neo-Lamarckian" notion being discussed.

In essence, the following excerpt exemplifies Harms' "neo-Lamarckian" philosophy (my translation)[3]:

> The more we immerse ourselves into the thinking of Darwin, which includes the non-vitalist Lamarckism, the more we recognize that it has also nowadays unlimited validity: the formation of species and with it evolution is in its course mechanistic, it is determined by the environment [...]. At first, it is of little importance whether the resulting, newly adapted form is only a permanent modification (somatic mutation) or a heritable one (idiomatic modification). Under correspondingly enduring and stable environmental conditions, the former will lead to genetic changes or the formation of new genes and radicals (Potthast and Hoßfeld 2010, p. 442).

Harms' attempt to reconcile the Lamarckian and Darwinian positions was quite clear in the above quote. In this sense, Harms fit the archetypal "neo-Lamarckians" in Hull's schema, in which "the neo-Lamarckians did not form a group but worked largely in isolation from each other. Instead of viewing themselves as renegades,

---

[3]I thank Raffael Himmelsbach for his help with the German translation.

they tended to see themselves as conservatives championing a wider view of evolution against the overly restricted position of the neo-Darwinians." (Hull 1984, p. 1). Harms saw the great Darwinism–Lamarckism divide in terms of relative differences rather than absolute contradiction. As he framed it, although the Darwinian theory of natural selection may never be discarded, Lamarckian theory was indispensable for explaining the emergence of functionally adaptive traits. Despite the fact that Harms was an adherent of the doctrine of "the inheritance of acquired characteristics" (*Vererbung erworbener Eigenschaften*), he was more articulate about the interplay of Darwinian and Lamarckian paradigms than about the exact mechanisms by which the acquired characteristics were inherited.

Like most neo-Lamarckians, Harms was a synthesizer who set himself the task of harmonizing seemingly contradictory systems. He wanted to bring the plausibility of environmental modification into the Darwinian program. Not just because Darwinian evolution was "cruel, wasteful, and opportunistic," but also because it ignored the possibility that evolution might not be as gradual as Darwin thought it was (Hull 1984, p. iv). Harms followed the Lamarckian convention to call upon at least considering environmental influences in addressing how species change rather than just natural selection (Burkhardt 1984). For Harms, recording adaptive responses to the environment and its philosophical implication for the Darwinian framework preceded answering specific questions in Lamarckian inheritance such as what the "*langandauernden und gleichmäßigen Umweltbedingungen*" were and how mutation can be organically passed on to the next generation.

Harms' take on the ontogeny-phylogeny relationship and his developmental thesis were inseparable from his Lamarckian worldview. For the purpose of illustration, I will briefly discuss two of his seminal works: cytoplasmic investigation of *Unionidae* development and the theory of germ cells.

The postembryonic developmental phases had always intrigued Harms, and he was attentive to recording the general conditions under which regeneration occurred. One of his best-known studies concerned the parasitic development of glochidia on *Unio* (Harms 1909). Glochidium was a parasitic larva feeding on freshwater mollusks. When attaching segments of glochidia to the host, the epithelium cells of the mussel around the glochidium began to proliferate. Harms suggested that the proliferation of epithelial cells was caused by the external implantation of glochidium upon the host cells. This was evident in the vigorous multiplication of epithelial cells just below the glochidium. Harms further described a slight growth of the shell of *Unio* as an indicator of external changes of the glochidia, although his interpretation was disputed by other researchers (Young 1911).

In *Körper und Keimzellen*, Harms discussed the origin of germ cells from the dynamics of post-embryonic differentiation and particularly the likelihood of regeneration of germ cells from somatic cells. He bypassed the Weismann-Roux "*Descendenztheorie*" which relied on the preformation of germ-plasm to explain regeneration. Instead, Harms drew upon his studies on the formation of germ cells of the opposite sex from the peritoneal epithelium (abdominal lining) to build an alternative theory of germ cells. He highlighted the acquired ability of regenerative capacities of germinal epithelium from which germ cells were derived. Noteworthy was the way in which he attributed the regeneration of germ cells to epigenetic manipulation (from transformed somatic cells) rather than pre-existing conditions (preformed germ-plasm). From Harms' perspective, the origin of germ cells was an affirmative example in favor of the environmental cause of genetic changes. It was changes in the organism's somatic environment that triggered the proliferation of epithelial cells that lined up the sex glands.

Does inheritance from changes of somatic environment count as an occasion of Lamarckian inheritance? David Hull wrestled with this question skillfully by demonstrating that it was semantics that was really at play. Here "semantics" was used to underline the careful use of terms and definitions for making necessary distinctions among closely related phenomena. By "Lamarckian inheritance," one could refer to the transcription of adaptive changes in the organism's body into hereditary materials and the subsequent passing on of these materials to the offspring. Or, one could consign "Lamarckian inheritance" to the surmise that phenotypical traits from one organism could transmit directly to another organism without touching upon any genetic materials. These were two very different ideas that were entangled under the "Lamarckian" rubric. Harms' proposition resembled the first variant more, but as Hull interpreted the issue, the crux here was not whether adaptive characters could be transmitted at all, but over the origin and the specificity of the transmission of those adaptive traits. Even if Harms made a convincing case that the "transformed somatic cells" could be transmitted to the germ cells which later transmuted the reproductive organs, he did not prove that the environment was the principal force for heredity and evolution because those changes in the somatic cells were very likely the result of a selection process rather than environmental induction. In plain language, our bodies and genes change over time as we adapt to the environment, but these adaptations cannot pass over to the next generation immediately. As a species, we can only change slowly and over a long course of natural selection. We modify gradually and somewhat accidentally, not inductively.

2.3 Harms and the Nazis

For my purposes, it is important to distinguish what Harms had achieved from what he left open. His objection to Weismann's presumed separation of germ cells and somatic cells proved prescient. Reproductive cells and hereditary materials were not independent from the rest of the body. Those who subscribed to a rigid specialization of cells would be taken aback by the later technology of cellular reprogramming. A case in point was the successful induction of pluripotent stem cells from adult somatic cells by a Japanese team. The lead scientist, Shinya Yamanaka, was awarded the 2012 Nobel Prize in Medicine.

What Harms did not do was to give an adequate elaboration of the mechanics for Lamarckian inheritance, let alone the relative contribution of ancestral, inherited, and adaptive factors in shaping ontogenetic development. No one, not even J.B.

Lamarck himself, has accounted for the mechanisms of inheritance of acquired characteristics in persuasive and conclusive details (Burkhardt 1984). Recently, some evolutionary biologists have argued that it was unfair to reduce the scope of Lamarck's thinking to merely the inheritance of acquired characteristics. Writing slightly before he died, Stephen Jay Gould (2002) championed a more broadly conceived landscape of Lamarck's evolutionary theory.

So why didn't Harms straighten out the record once and for all? Why didn't he go ahead and lay out the mechanisms of Lamarckian inheritance? To address these questions, it is helpful to turn to David Hull, who has tried to explain why the Lamarckian theory of heredity and evolution was difficult to illustrate experimentally. It was not that Lamarckians were less scientifically rigorous than Darwinians, but to sufficiently demonstrate the vast environmental impacts on short-lived organisms was very difficult, if not inconceivable. Environmental variations are simply too immense and enduring to be internalized by the genetic capacity of an organism within a limited life span. But demonstrable falsifiability was not the main reason for the widespread unpopularity of Lamarckism among evolutionary biologists. Both David Hull and Richard Burkhardt have pointed out that the unjust condemnation Lamarckian advocates received had more to do with the power dynamics and social divides in the scientific community than Lamarckism's empirical deficiency. Darwinian champions won along with Mendelian genetics and the discovery of DNA while Lamarckian followers lost with the Kammerer and Lysenko affairs. Scientific controversy was never just about science; it was also about the social and political phenomena in the wider culture and how people treated each other.

As it turned out, Lamarckism was not just a disreputable position among scientists, it was perceived as a political "crime" in the eyes of the beholder. Harms' "neo-Lamarckian" way of thinking was to cause him troubles with the bureaucrats of the Third Reich.

In the winter of 1935, Harms was appointed to the renowned position of "Haeckel Chair" of the department of zoology at the University of Jena. Connected to the directorate of the Zoological Institute at Jena was that of the Phylogenetic Museum. For a while, Harms enjoyed dual positions of privilege. But Harms' Lamarckian view clashed with the ideology of the Nazis. In 1938, Harms was relieved from his directorate of the Phylogenetic Museum because of his Lamarckian beliefs. Harms' antagonistic relationship with the political authority intensified when his former assistant at Tübingen—Gerhard Heberer—was appointed to the directorship of the Phylogenetic Museum at the behest of SS (*Schutzstaffel*) commander Heinrich Himmler. As Heberer was promoted to SS-First Lieutenant (*SS-Obersturmführer*), Harms was forced to resign from his Haeckel professorship. Harms was rehabilitated after World War II in 1946 and became the rector of the University of Jena but his petition to return to the Zoology Institute at Tübingen was denied. As Potthast and Hoßfeld (2010) lamented, zoology at Tübingen was in free-fall afterwards as the scientific team was squeezed out by the Nazis cliques. Both now and then, scientific conviction has never been completely isolated from political intervention. Harms' case was no exception.

The biographical profile of Harms is explored at length here because of the significant impact of Harms on Bei's scholarly development. It is impossible to grasp the host of scientific and political issues that puzzled Bei throughout his life without immersing ourselves first into Harms' intellectual world. To assess Harms' lasting influence on Bei, we now turn to Bei Shizhang and his theory of cell reformation.

2.4 Bei Shizhang and His Theory of Cell Reformation

Despite much respect to Bei Shizhang's lifelong dedication to furthering the development of scientific enterprise in modern China, academic renown was still missing from his list of accomplishments. Bei's biggest regret was that he didn't succeed in advancing the cell reformation theory that he proposed in the early 1940s. After more than seventy years, Bei's attempt at giving an alternative interpretation of cell differentiation is still met with resistance from the mainstream scientific community. The lack of international recognition of Bei's scientific achievement led some Chinese scientists to mock him as a "centenarian" (百岁老人), insinuating that Bei was just an old codger without much scientific credibility other than his longevity.

I will first describe the content of his theory of cell reformation, followed by explanations on what motivate his research inquiry. These background understandings, in my opinion, are helpful for making sense of the controversy surrounding his theory.

Bei's theory of cell reformation casts doubt on the prevailing cytological paradigm exemplified by the Virchowian motto *omnis cellula a cellula* (all cells come from pre-existing cells) by drawing on the result of his empirical study of the developmental morphology of an autochthonous prawn in China.

The story of Bei's iconoclastic research typically took us back to 1932 (Bei 1992). The official story began with his finding of an intersex strain of an arthropod in the swampy field of Songmuchang on the outskirts of Hangzhou. Indigenous to the Chinese southeastern shores, its unusual reproductive process captured the curiosity of this German-returned zoologist. *Chirocephalus nankinensis* (南京丰年虫),[4] as it was called, was a fascinating shrimp for Bei because of its unusual mechanism of cellular proliferation during sex change. The sample breed under his investigation was of hermaphrodite nature, setting *c. nankinensis* apart from other sexually dimorphic arthropods. Microscopic observations showed that characteristics of both sexes co-existed in a typical *c. nankinensis* strain. Although *c. nankinensis* combined

---

[4]The English scientific name was coined to capture the fact that this insect is a Nanjing (aka Nanking)-originated species (*nankinensis*) in the family of *chirocephalus*. The Chinese name was given by a folklore belief that the appearance of this shrimp in wintertime, along with heavy snowstorms, is a harbinger of a productive year as summarized in the Chinese proverb "heavy snow forecasts a good harvest year" (瑞雪兆丰年).

within one organism both male and female characteristics, the distribution and proportion of these sex features were not uniform among all types. According to the relative sexual traits from the specimens he collected, Bei divided them into male and female intersex types. He noted that at certain developmental stages, both sexes underwent sex reversal. The female intersex transformed into male and male into female. He further sub-divided the female intersex into weak, middle, and normal female intersexes while the male intersex was divided into weak and normal male intersexes by virtue of their secondary morphological features. Thus, there were altogether five types of intersex strains of *c. nankinensis*. When the weak male intersex (弱勢雄原中间性) underwent gonad reversal, the germ cells disaggregated their cellular contents into yolk granules or substances similar to yolk—a process which he called "cell deformation" (细胞解形)—then re-aggregated into an adult cell incrementally from these chromatin-bearing entities during the transformation of germ cells from weak female intersex (弱勢雌原中间性) to weak male intersex. Bei argued that the yolk granules outside the cytoplasm of the oocytes "reformed" from yolk granules in this manner, which he called "cell reformation." The new germ cells were formed not by cytokinesis but rather generated from the corpuscular substances, i.e. the yolk granules. His conclusion was that it was possible for cells to reproduce by means other than cell division. "Cell reformation" was thereafter hailed as an alternative way by which cells multiplied.

The above episode was probably the most well-known part of Bei's work on the theory of cell reformation. It was published in a Chinese reminiscence article by Bei himself in 2003, nearly seventy years after he had first conducted the research (Bei 2003a). After Bei's death, his long-time assistant and one of his vocal supporters—Wang Guyan wrote a short English sketch in remembrance of Bei's cell reformation theory. Published in 2010, Wang's cameo stood as the only existing writing in the Anglophone community about Bei Shizhang and his work on cell reformation (Wang 2010a, b). A lack of publications in English journals was a blot on Bei's escutcheon. Those who denounced Bei and the theory of cell reformation held that if his theory was really credible, it should have appeared in either *Nature* or *Science*, if not both. The absence of an imprimatur from the mainstream Anglo-American authorities was iterated as indicative of the inferior quality of his work. One whistle-blower who styled himself as the "science cop" in China went as far as calling the theory of cell reformation "pseudoscience" and Bei a "third-rate scientist," an accusation that I will explore later.

My purpose here is not to rehabilitate Bei's theory or reputation, but I think it is necessary to consider the entirety of his work before jumping to any conclusions. For better or worse, the familiar story was an over-simplified tale of a man with a complex and eventful life. The official saga portrayed Bei as a genius who challenged a cherished foreign theory with an indigenous Chinese organism. Bei and his admirers also perpetuated this myth of a scientific hero by emphasizing his ambition to undercut existing views on cell differentiation. Ever since Bei wrote the first research article on *c. nankinensis* in German, he had relentlessly cast his theory as an alternative to the traditional viewpoint held by Rudolph Virchow. He was determined to refute Virchow's claim that cell division was the only way by which new cells generated by upholding his cell reformation theory as undermining the *philosophia perennis* in cell biology (Bei [1943a], [1943b] 1992).

Among the many important questions obscured by this abridged story was what inspired Bei to pursue an alternative cell theory. His eureka discovery of *c. nankinensis* in Hangzhou was one thing, yet how to make sense of what he saw under the microscope was another. It was not just a matter of instrumental conditions. Translating empirical findings into a theory did not follow a linear pattern because theorizing cellular activities required an agenda to determine the proper relationship among cellular and subcellular organelles. It entailed much more than the ability to optimize the quality of pictures and the extent of visualization. With advanced microscopes and the right techniques of preparation, a set of working assumptions is required in order to interpret the biological phenomena in a petri dish. In other words, what matters most is not how much one can see, but where to look for what one assumes is being seen and how to decide what to make out of it. In Bei's study, why did he focus on the sexual metamorphisms and not other structural and functional aspects of *c. nankinensis*? What drove him to divide *c. nankinensis* into five categories? Where did he get the ideas of "deformation" and "reformation?" In short, what led him to such a particular cellular interpretation?

The *naturphilosophie* of Wilhelm Harms, which informs his research on the relationship between the endocrinal secretions of sex glands and the whole organisms, holds the key to the above questions. Bei's formulation of the cell reformation theory reflects the developmental worldview espoused by Harms (1939).

According to Ying Youmei, Harms' emphasis on the entire developmental cycle made a permanent impression on Bei. To probe into the regenerative power of the labile tissues and organs such as the primordial germ cells, Harms divided the life cycle of a typical organism into three stages: *progressive periode*, *stationäre zustand*, and *regressive periode*. Ying remarked that "Bei appreciated Harms' philosophy of addressing a problem comprehensively from a developmental point of view." (Ying 1992, p. 5)

Ying's analysis of Harms' influence on Bei's intellectual evolution captures the correlation between Harms' developmental philosophy and Bei's framework on cell reformation. Harms stressed sexual physiology and regeneration of germ cells whereas Bei studied morphological changes of germ cells during sexual metamorphosis; Harms looked at growth and developmental processes whereas Bei examined the cytoplasmic changes during cell differentiation. More strikingly, Bei adopted Harms' terms to illustrate the core concepts in the theory of cell reformation. Bei attributed the transition from male to female intersex to the "*regressive periode*" (退行性) while the female to male intersex to the "*progressive periode*" (前进性). Not only were "regressive period" and "progressive period" a direct importation of Harms' language; the ideas were the primary conceptual grounds for building Bei's theory. The "regressive period" corresponds to the "deformation" (解体) of germ cells and the "progressive period" to the "reformation" (形成) of reproductive cells. Harms' work was the key to unlock Bei's philosophical foundation.

Even before Bei postulated the theory of cell reformation, Harms' scientific style already colored the topic, driving questions, and analysis of Bei's doctoral thesis. Using the nematode *anguillula aceti* (a type of vinegar eel) as the research material, Bei (1928) embarked his doctoral research on an "experimental–morphological investigations on nematodes" (*experimentell-morphologische untersuchungen an nematoden*). He proposed to study the complete life cycle rather than just the embryonic stage. One of the objectives of his dissertation was "to follow the entire life cycle" (*den gesamten Lebenscyclus zu verfolgen*) of the nematode. His analysis of the development of *a. aceti* followed the idioms of *progressive, stationäre*, and *regressive phasen* introduced by Harms. In addition, Harms' abiding interest in sex determination and germ layers left an indelible mark on Bei's discussion of the findings. Bei's dissertation incorporated a number of handwritten drawings on "the growth of sex cells and germ cells" (*wachstum der geschlechtszellen* and *keimzellen*). "Germ-line and the determination of somatic cells in the formation of the germ layers" (*keimbahn und die determination der somazellen bei der bildung der keimblätter*) was a major theme in Bei's mapping of the morphological differentiation of *a. aceti*.

In short, the reason Bei focused on sexual metamorphisms rather than other physiological aspects of *c. nankinensis* is a legacy of Harms' enduring interest in sex transformation and reassignment. Harms' tripartite analytical structure also exhibited a discernable influence on Bei's framework and his choice of words in categorizing the intersex strains of *c. nankinensis*. Harms and Bei had a lot in common as both were attracted to the developmental mechanics of living organisms, with a special emphasis on the differentiation of germ cells.

The intellectual continuity between Harms and Bei is important for considering the kind of oppositions against Bei. The objectionable nature of neo-Lamarckian cause of inheritance is an underlying factor, but the controversy of the theory of cell reformation is complicated by the political interests revolving around the issue of cell origin.

2.5 Bei Shizhang, O.B. Lepeshinskaya, and the Controversy of Cell Theory

On 24 August 1964, during his meeting with the Japanese physicist Shoichi Sakata in the company of Chinese physicist Zhou Peiyuan, Mao Zedong enunciated his intellectual interest in the origins of cells: "We should study the origins of cells. The cell has its nucleus, a mass of protoplasm, and a membrane. The cell is organic, and so there must have been non-cellular forms (cytooes) before there was the cell. What was there before the cell was formed? How was the non-cellular form changed into the cell?" (Mao 1964).

Apparently, his interest in this biological area was partly triggered by the work of a Soviet biologist: "there is a woman scientist in the Soviet Union who has been studying this problem, but no result has been reported." (Mao 1964). The "woman scientist in the Soviet Union" in Mao's speech was Olga Borisovna Lepeshinskaya. At that time, Mao probably did not know that before his declaration of interests and even before China was governed by his regime, Bei Shizhang was already intrigued by Lepeshinskaya's work.

In 1943, Bei published a Chinese research paper entitled "Yolk Granules and the Reformation of the Cell" (卵黄粒与细胞之重建). It was contained in the inaugural issue of the Chinese journal *Science*. It was also the first place in which his theory of "cell reformation" was introduced to the scientific community to explain the process of cell differentiation at a time when the world was set ablaze by the flames of WWII.

Bei's 1943 paper started off by reviewing the research efforts made by the Soviet biologist O.B. Lepeshinskaya between 1936 and 1937. Her name is now infamous among historians of science, and for good reasons. In May 1950, O.B. Lepeshinskaya rose to superstardom in Stalinist Russia with her "new cell theory." As an anti-Virchowian, Lepeshinskaya claimed to have proved that the basic properties of an organism were contained not in cells but in some amorphous "vital substance." It was later revealed that the alleged bearer of all vital processes and materials for producing cells—"vital substance"—was founded upon Stalin's blessing rather than replicable results. Her reputation was completely destroyed as her experimental results failed to stand the test of other scientists (Zhinkin and Mikhailov 1958). Her coronation as a famous scientist by her receipt of a Stalin Prize and election to the Soviet Academy of Medical Sciences were glacially described by a former Soviet physician as representative of "a carefully staged farce of collective ecstasy for the 'great discovery.'" (Rapoport 1991, p. 266)

Although Lepeshinskaya's meteoric rise did not last long, the political reper-cussions were more far-reaching. Her "inglorious demise" was usually cited not as a reflection of her individual character but as a general problem inherent in the philosophy of "dialectical materialism." For instance, the exposé article from *Science*, which revealed the irreplicable results of various sorts reported by O.B. Lepeshinskaya, began with the characterization of her theory as "a new dialectical-materialist cell theory." (Zhinkin and Mikhailov 1958, p. 182). Loren Graham, the award-winning historian of Soviet science and philosophy, also denounced Lepeshinskaya's cell theory scornfully, but Graham did not regard her theory as discrediting dialectical materialism writ large. Graham considered Lepeshinskaya an unbridled careerist who rigged the political system for her own gains. But it was her lack of personal integrity, not dialectical materialism as a philosophy of science that was to blame. Beneath her character flaws, Graham suggested that there was very little connection between her cell theory and the intellectual content of dia-lectical materialism (Graham 1974).

The denigration of Lepeshinskaya forms the backdrop against which we can assess Bei and his reputation because of his close affinity to Lepeshinskaya. As a matter of fact, Lepeshinskaya was Bei's Soviet analogue: Both were prominent biologists living at the revolutionary heights in their home countries (Bei during Mao's regime and Lepeshinskaya during Stalin's). Both came up with alternative interpretations on cell proliferation and had contemplated on the question of the origin of life. Both wrote papers in German in addition to their native languages. They met twice in Moscow to exchange views and papers. Since they were acquaintances, and since Lepeshinskaya was a sloppy charlatan, for many observers it followed that Bei was probably a fraudulent scientist too.

It is worth knowing that the correspondence of Bei and Lepeshinskaya ran deeper than their social acquaintance and comparable theoretical contentions. Bei was intrigued by Lepeshinskaya's work before China turned communist and before Lepeshinskaya became one of Stalin's favorites. Prior to Lepeshinskaya's political debut in 1950, Bei had already studied and made reference to her research papers. This brings us back to Bei's 1942 paper which first appeared in the Chinese magazine *Science.* The papers Bei had access to were written in German and published in the Japanese journal, *Cytologia* in 1936 and 1937. Considering the chronology of events, it was highly unlikely that Bei paid attention to Lepeshinskaya because he was yielding to political pressures. If Bei was a "Chinese Lepeshinskaya," the parallel was not political service to the parties—led by the Bolsheviks or the "Chinese Bolsheviks." (Luk 1990). The facts call into question the stereotypical thinking that lumps Bei and Lepeshinskaya together simply because they were both biologists in the Red East.

Nor was Bei an uncritical supporter of Lepeshinskaya. Bei was aware of the shortcomings in her papers. What drew Bei's attention was Lepeshinskaya's conjecture of "the formation of cells from yolk spheres in chicken embryos" (*bildung von zellen aus dotterkugeln beim Hühnerembryo*). Bei was attracted by her revolutionary thesis that yolk spheres could generate cells after "remaking" (改塑). But at the same time he cautioned that Lepeshinskaya's discovery exceeded the general principle of cell generation, and thus in need of a careful examination (Bei 1943a, p. 111). Bei's skepticism about Lepeshinskaya appeared in print before Lepeshinskaya was formally ostracized by the Western scientific community. Although Bei did not disprove Lepeshinskaya's claims, he did not unconditionally accept her arguments either. His mindfulness and independent thinking had more to do with his fascination with cell studies rather than path-dependency upon the West.

Bei's interest in and skepticism of Lepeshinskaya prior to her political fame and scientific infamy should not be underrated. The cytological connection between Bei and Lepeshinskaya had been pre-fabricated before the arrival of political intervention from Kremlin and organized skepticism from America. It was not political struggles or scientific networking that cut across the Bei–Lepeshinskaya line. Rather, it was a shared intellectual curiosity in the past and future of cells that was the common thread. However, the common interests in examining the development of cells in relations to their surrounding fluids should not conceal their differences. Not only was Bei skeptical of Lepeshinskaya's paradigm-shattering contention; he held a different interpretation of the cellular phenomenon under investigation. Bei also disagreed with the specific use of terminology in Lepeshinskaya's papers (my translation):

Lepeshinskaya's view that yolk spheres from chicken embryos could be remade into cells, in my speculation, was the well-nigh reformation of yolk spheres. As the meanings of remaking and reformation differ. Remaking means creation from anew, but reformation simply means revitalization. The latter must be easier to accomplish as the ingredients were all there (as inherited) from the historical background (Bei 1943a, p. 115).

The research materials of Lepeshinskaya's 1936 and 1937 papers differ quite significantly from these of Bei's 1942 study. But the meaning of the Bei–Lepeshinskaya interface, at least from Bei's point of view, was the connotation and interpretation of the keywords. He interpreted Lepeshinskaya's investigation of "*die entstehung von zellen aus dotterkugeln*" as "remaking" cells from anew. The "remaking" (改塑) of new cells from yolk spheres in chicken embryos created an image of the de novo generation of cells which undercut the historical continuity of cells. Bei was unsure of this "ahistorical" implication signified by the word "改塑." He did not overrule Lepeshinskaya's choice of the word, but he was confident that the word "reformation" (重建) was more appropriate. Although there were other English equivalences to the Chinese word "改塑" such as remodel, reconfigure, or recreate, the bone of contention was the historicity of cells. Bei was in favor of seeking words that would preserve the undertone of the historical continuity of cells. He was concerned that a complete severing from the genetic past would make regeneration difficult to occur. Reforming cells were more plausible and believable than remaking cells. Semantics was at the center of Bei–Lepeshinskaya scholastic exchange.

Bei's concern with the choice of word was wise, as the issue would become the center of disagreement in contemporary debate surrounding his objectionable cell theory. On 8 November 2005, a Chinese blogger sent an excerpt of Bei's cell reformation theory to Fang Zhouzi (方舟子), who is known for writing science popularization literature and exposing academic misconduct in China (*Nature* 2012). The sender was hoping that Fang would publicly denounce Bei's theory as fraudulent. Fang re-posted the excerpt to the website http://www.xys.org but did not add further comment.

Registered and based in the US, http://www.xys.org (the URL was derived from its Roman initials of its Chinese name 新语丝, *The New Threads*) is a Chinese website launched and maintained by Fang. The website provides mainland and overseas Chinese a virtual space to share and discuss topics related to science and society in China. After the post appeared on *The New Threads*, a web user picked up the excerpt and waged an online war with Fang over the credibility of Bei's theory. Under the alias *Tomoe*, this user did not regard Bei's cell reformation theory as bogus. *Tomoe* argued that some aspects in Bei's theory were consistent with the embryonic development in *Drosophila*. *Tomoe* pointed out that the embryos of *Drosophila* do not undergo transcription and cytokinesis in the first two hours but the nucleus divides once every nine minutes and results in hundreds of nuclei after the ninth division. The nuclear DNA comes from cell division but the nucleus constituents such as proteins and nuclear membranes come from material deposit. The enzymes, energy, and nucleic acids necessary for replicating DNA also come from the mother cells. The newly formed cells are built on the basis of cytoplasm from these mother cells. The phenomenon is consistent with one of the tenets in Bei's cell reformation theory, which states that new cells are generated on the basis of existing materials from old cells. *Tomoe* suggested that "a special form of division" would be a more appropriate term to describe this mechanism rather than "cell reformation." But whatever it is called, *Tomoe* claimed that there is a considerable overlap between the developmental embryology of *Drosophila* and Bei's theoretical assertion. What *Tomoe* objected to was a premature and sweeping denunciation of Bei's cell reformation theory.

Tomoe benchmarked Bei's theory against external criteria in developmental genetics; he did not uncritically glorify Bei, but he found Bei's vision both prescient and precious as he stated, "What I admire is the phenomenon he has observed some seventy years ago" and concluded that "his theory is now common sense, but it was a breakthrough at that time." Yet Fang and his followers were not ready to grant Bei this special acclaim of prophetic insight. Fang maintained, "All, not just some, viewpoints of old Bei are 'mistaken'." Fang regarded Bei as a vitalist and interpreted Bei's theory as advocating the de novo generation of new cells by denying the historical continuity of cells: "regardless of what materials are utilized, 'cell reformation' denies the genetic continuity of cells. It suggests that environmental condition alone (without the need of genetic order of the old nucleus) is enough to generate new cells and nucleus from scratch…" (*The New Threads*, 2005)

At the heart of this debate lies the question regarding the historical continuity of cells. Fang rejected Bei's theory on the grounds that cell reformation defies the historical continuity of cells, in contrast to *Tomoe* who suggested otherwise. For Fang, what was at stake was not Bei's experimental sophistication or his choice of research substrates; nor did Fang display much interest in the cytomechanical similarities between *Drosophila* and *c. nankinensis*. Fang was skeptical of Bei's theory because he considered the theory of cell reformation a heresy against the doctrine of cell continuity.

The paradigm of cell continuity is a core theme that has run through much of the history of biology in twentieth-century China. Scientists have been concerned with the lineage of cells ever since Virchow, if not earlier. Western scientists had been preoccupied with the question of how new cells were formed and where cells came from ever since the establishment of cells as the fundamental unit of life (Harris 1999). In Republican China, Bei dedicated much time and energy into investigating the formation, constitution, and interpretation of how cells arose. By advocating a cell-from-yolk cause of cell generation, Bei was among many scientists of the world who offered a cell-from-X hypothesis in the early twentieth century. Specifically, Bei Shizhang postulated "reformation" as a theoretical explanation of some of the transformation of cells. While once sympathetic to Lepeshinskaya's cytological theory, Bei was insistent upon choosing the right term in order to not sound ahistorical.

Under Mao, the "origins of life" were placed in the framework of dialectical materialism. To demonstrate relevance of his work to state-approved agendas at the height of the Cultural Revolution, Bei aligned his theory on cell lineage to Mao's known interest in cell origins. What Mao offered was a justification for Mao-era scientists to work on issues related to cell lineage. In the post-Mao era, science writers and informed netizens shared the same concern over the historicity of cells. Both before and after Mao, Chinese scientists and critics have cared much more about the concept of maintaining biological continuity with the past than with the exact mechanisms of how this was achieved.

Was Bei's study a Chinese scam that resembled the Soviet-style pseudo-science presented by O.B. Lepeshinskaya? This question was contemplated by Hu Wengeng (胡文耕), one of the most widely-cited philosophers of biology in contemporary China. In a chapter devoted to the topic "cell origins and cell reformation," he dealt with the objection that Bei's work was dubious simply because it looked too much like Lepeshinskaya's. A popular attitude held that since cell reformation was similar to Lepeshinskaya's work, and since she failed, therefore it was difficult for the work of cell reformation to obtain positive results. Hu's tactic was to highlight the differences between Bei and Lepeshinskaya in terms of research materials, theoretical emphases, and experimental conditions. Hu's point was that although there were some overlaps between Bei and Lepeshinskaya, their differences should not be overlooked. Their dissimilarities were more substantive and suggestive than their superficial similarities. For example, one of the marked Lepeshinskaya–Bei discontinuities, according to Hu, was that Lepeshinskaya purported to completely replace Virchowian paradigm with her "vital" theory while Bei was merely offering an alternative way to broaden the cytoscape dominated by Virchow. Moreover, Bei did not commence his study of the developmental cycles of c. nankinensis with the intention to root out advocates of Virchowianism and Weismanism the way Lepeshinskaya did. He was not a political demagogue manipulating science for his personal gain. Unlike Lepeshinskaya, Bei did not start from a shaky foundation that rested on an unempirical, quasi-supernatural belief in "vital substances." He never claimed that cells, or any fundamental units of life, could come from mysterious non-living matter a priori. He never wanted to dismiss rigorous evidence in favor of an inexplicable vitalistic cause of life (Hu 1982).

As far as I know, Hu's vindication is probably the only existing writing that appraised the theory of cell reformation from a philosophical perspective. Hu addressed "the possibility of the theory of cell reformation" in a historiography paper (Zhang 1996). Situating the theoretical plausibility of the theory of cell reformation within the context of the natural history of cells, Hu concluded that Bei's theory should not be dismissed of its philosophical plausibility simply because it looked similar to the failed attempt of Lepeshinskaya. Hu called for further experimental corroboration to clarify unclear points in Bei's theory.

While Hu gave Bei's theory the benefit of the doubt, not all Chinese reviewers shared Hu's sympathy. The Chinese neurobiologist, Rao Yi (饶毅), recently expressed his distrust of Bei and his work. Rao was trained at Harvard and UCSF. He had worked at the School of Medicine at Washington University in St. Louis before serving as the head of a scientific institute at Northwestern University. In 2010, his high-profile decision to give up not just his tenured professorship but also his American citizenship to return to China put him in the media spotlight

(FlorCruz 2010; LaFraniere 2010). He is now the dean and chair professor of the School of Life Sciences at Peking University.

Rao (2009) authored a collection of essays expressing his views on a number of popular issues surrounding science and science management in China. The range of topics appearing on his narrative radar included backyard stories of Marie Curie, gender inequity in science, tips on how to read scientific literature, lists of scientists whom he thought deserve Nobel accolades for their achievements but hadn't won yet (many of the scientists on his list later won the Nobel awards, making him a stunningly accurate predictor of future Nobel prize-winners). The point is, Rao's compilation was not as philosophically oriented or intellectual minded as Hu's monograph. Rao's anthology was intended for general Chinese readers rather than philosophers of science. One of the essays was a short eulogy he wrote to commemorate the late biochemist Zou Chenglu (邹承鲁) when Zou died in 2007. Entitled "Zou Chenglu: The Good People Loved Him, and the Bad Hated Him," Rao walked through the major milestones in Zou's righteous life while venting his frustration and grievances against a certain "centenarian" and his theory of cell reformation:

> The "centenarian" has made contribution to the development of Chinese science. However, compared to his contemporaries such as Feng Depei in physiology and Wang Yinglai in biochemistry, the quality of his academic achievement was less outstanding, and even quite below that of his contemporaries'. When I first came to visit the newly established research center of his, I was brought to the exhibition room in which his "cell reformation" results were on display. I said his research was inappropriate.....It was inappropriate as anyone can take a gander and see his limited "achievement".....When I expressed my concern, I did not know that academician Zou Chenglu had raised similar objections in the past and had paralyzed their interpersonal relationships...Perhaps because the issue is now timeworn, I have not been reprimanded. But things could not have been easy for Zou as Zou used to work under him...Those who did not know about developmental biology might regard Zou's objection against him as groundless. Reportedly Zou said his research in China was quite inactive over the years. He seldom published papers. Readers can check out bibliography to evaluate Zou's comment on their own. There is no need for other people to explain (Rao 2009, p. 159).

When Rao's article was first published in 2007 in a Chinese periodical, Bei Shizhang was still alive. This might explain Rao's indirect reference to him as the "centenarian" rather than spelling out his name. Yet there is no doubt that Rao was talking about Bei Shizhang. Besides invoking Bei's personal trademark—"cell reformation"—Rao's speculation on the glacial relationship between Bei and Zou was situated when Zou Chenglu was a biochemist at the Institute of Biophysics, thus explaining why Zou was "uneasy" when he worked with Bei at the same institute. Also, Feng Depei and Wang Yinglai were both scientists in Bei's cohort. All the signs point to Bei as the target figure in Rao's essay. If we accept that Bei Shizhang was indeed the target of allusion in Rao's essay, the next question is: why did such a young and brilliant scientist choose to publicly denounce Bei?

One of the main reasons for a lack of approbation of Bei's theory of cell reformation among the mainstream scientific community is his insufficient explication of details. What Bei called "cell reformation" encompassed primarily the principle that cells could be "reformed" from the yolk granules that enveloped the cells under suitable conditions. But "deform," "reform" are vague terms in a scientific study, nor is the phrase "suitable condition" specified with enough details. The biggest problem is that even if one observes an abnormal pattern of cell differentiation in *c. nankinensis*, is it justified to generalize from this one isolated occurrence to the overall cytoscape? Even more so, since Mao's declaration of interest in the origins of cells, Bei began to sell his theory as offering a potential explanation for cell origins and even the evolution of life. The problem is that issues as complicated as the origins of cells are very difficult, if not impossible, to illustrate in an experimental setting. It is not that Bei was less attentive to scientific rigor but that sufficiently demonstrating the vast array of environmental and hereditary factors on short-lived organisms in a laboratory was almost inconceivable.[5]

In sum, cell reformation theory is hardly an undisputed set of ideas. The controversy has its root in its Lamarckian view of life, of which Wilhelm Harms was an adherent. Even though Harms was sympathetic to the Lamarckian cause of inheritance, the truth is that he did not give sufficient details and evidence to support the Lamarckian interpretation. Harms' lifelong emphasis on the environmental conditions and organic development shaped Bei's formulation of scientific theory. Apparently, this "intellectual trait" of dedication to knowledge synthesis but failure to account for internal mechanics also inadvertently passed on to Bei. On the one hand, Bei was cultivated to synthesize specialties with divergent methodological approaches and theoretical worldviews. On the other hand, Bei's humble academic record also stems from the same intellectual commitment: he saw the big picture and the connection among previously separated fields, but did not attach enough weight to details and intricacies, and the details are important, especially when attempting to communicate the merits of a new theory to detail-oriented people.

2.6 Concluding Remarks

This chapter gives an introduction to the early life and academic disputation of Bei Shizhang—the founding father of biophysics in China. Controversy over Bei's theory is of direct relevance to the historiographical character of biophysics as the theoretical viewpoint upheld by a disciplinary founder is indicative of its epistemic

---

[5]Between 1988 and 2003, two Chinese volumes by the title of *Cell reformation I and II* were published. The twin volumes contain follow-up and new research on cell reformation under the laboratory guidance of Bei Shizhang. In addition to cell reformation of *c. nankinensis* yolk granules, sources of experimental data in these studies were broadened to include self-assembly of reforming cells in the early development of chicken embryos, nucleus reformation of cultured bone marrow cells of adult mice, cell proliferation of *chlamydia trachomatis*, and reconstruction of cells in *Rhizobium japonicum* in *glycine gracilis*. However, none of these experiments have been externally reviewed and/or testified by a third party. The figures and data published in these volumes remain less than satisfactory. See Bei, ed. (1988, 2003b).

content. The extent to which Bei's conception of biophysics is embodied in his theoretical outlook behooves us to study their correlation and broader impacts.

To investigate the formation of Bei's scientific worldview, I chronicled Bei's early schooling that led to his graduate study with Wilhelm Harms in Germany. From there, I explored the scholarship of Harms with the intention of highlighting Harms' intellectual influence on Bei's academic pursuits, in both a positive and negative light. I suggest that the disagreement on the credibility of the theory of cell reformation was ensnared in the larger sets of debate involving Soviet-styled biology, the perception of Marxist philosophy, and political patronage of cytology in twentieth-century Communist states.

The story of Harms and Bei stayed mostly in Germany. Since Bei returned to China in 1931, his cytological research would lie dormant for the next 30 years, during which he put the state-assigned administrative and organizational tasks before his own academic agenda. In the post-1949 period, Bei would have to put his scholarly interests on hold, as his duties would compel him to act as a discipline builder that would ultimately define him as a "founding father" of Biophysics in China. It is to these institutional building efforts we now turn.

References

Bei S (Sitsan Pai) (1928) Die Phasen des Lebenscyclus der Anguillula aceti Ehrbgund ihre experimentell-morphologische Beeinflussung. Zeitschrift für wissenschaftliche Zoologie 131:294–344

Bei S (Sitsan Pai) [1942] (1992) Diploide Intersexen Bei *Chirocephalus Nankinensis*. Science Record, reprinted in Selected writings of Bei Shizhang (SWB), pp 99–100

Bei S (Sitsan Pai) [1943a] (1992) The cellular reformation of yolk granules (卵黄粒与细胞之重建). Ke Xue, reprinted in SWB, pp 110–121

Bei S (Sitsan Pai) [1943b] (1992) Ueber die Transformation der Genitalzellen bei den *Chirocephalus*-Intersexen. Ke Xue, reprinted in SWB, pp 122–123

Bei S (Sitsan Pai) (ed) (1988) Cell reformation I (细胞重建第一集). Science Press, Beijing

Bei S (Sitsan Pai) (1992) Selected writings of Bei Shizhang (SWB) (贝时璋文选). Zhejiang Science and Technology Press, Hangzhou

Bei S (Sitsan Pai) (2003a) Seventy-year research on cell reformation (七十年的细胞重建研究). PIBB 30(5):I–VII

Bei S (Sitsan Pai) (ed) (2003b) Cell reformation II (细胞重建第二集). Science Press, Beijing

Burkhardt R Jr (1984) The zoological philosophy of J. B. Lamarck. In: Lamarck JB (ed) Zoological philosophy: an exposition with regard to the natural history of animals, The University of Chicago Press, Chicago, pp xv–xxxix

China News (2009) Various sectors in China wave goodbyes to the oldest academician Bei Shizhang (中国各界送别中院最年长院士贝时璋). China News, 4 November

Committee of Funeral Service for Bei Shizhang (贝时璋先生治丧委员会) (2009) Mourning for Bei Shizhang: an obituary (沉痛悼念贝时璋先生: 讣告), 29 October

Conklin EG (1897) The embryology of Crepidula: a contribution to the cell lineage and early development of some marine gasteropods. J Morphol 13(1):1–226. http://archive.org/stream/embryologycrepi00conkgoog/. Accessed 21 Dec 2014

FlorCruz J (2010) 'Sea Turtles' reverse China's brain drain, CNN World, 28 October

Graham L (1974) Science and philosophy in the Soviet Union. Vintage Books, New York

Gould SJ (2002) The structure of evolutionary theory. Belknap Harvard Press, Cambridge

Harms W (1909) Postembryonale Entwicklungsgeschichte der Unioniden. Zoologische Jahrbücher: Abteilung für Anatomie und Ontogenie der Tiere 28:325–386

Harms W (1926) Körper und Keimzellen vol 1 & 2. Springer, Berlin

Harms W (1939) Ders.: Lamarckismus und Darwinismus als historische Theorien. Ein Kampf um Überlebtes. Zeitschrift für Medizin und Naturwissenschaft 73:1–27

Harris H (1999) The birth of the cell. Yale University Press, New Haven

Hu WG (胡文耕) (1982) Philosophical questions in molecular biology (分子生物学中的哲学问题). Tianjin Renmin Press, Tianjin

Hull D (1984) Lamarck among the anglos. In: Lamarck JB (ed) Zoological philosophy: an exposition with regard to the natural history of animals. The University of Chicago Press, Chicago, pp xl–lxvi

Institute of Biophysics, Chinese Academy of Sciences (IBP-CAS) (2003) Bei Shizhang and biophysics: a festschrift (《贝时璋与生物物理学》纪念文集). http://www.ibp.cas.cn/sqzj/45zn/qdxlcbw/jnwc/. Accessed 21 Dec 2014

LaFraniere S (2010) Fighting trend, China is luring scientists home. The New York Times, 6 January

Lepeshinskaya OB (1936) Zur Frage nach der Neubildung von Zellen im tierischen Organismus. 1. Bildung von Zellen und Blutinseln aus Dotterkugeln beim Hühnerembryo. Hnerembryo. Cytologia 7:54–81

Lepeshinskaya OB (1937) Zur Frage nach der Neubildung von Zellen im tierischen Organismus. 2. Mitteilung Neuere Ergebnisse über die Bildung von Zellen und Blutinseln aus den Dotterkugeln des Hühnerembryos, hnerembryos. Cytologia 8:15–39

Luk MYL (1990) The origins of Chinese Bolshevism: an ideology in the making, 1920–1928. Oxford University Press, Oxford

Maienschein J (1978) Cell lineage, ancestral reminiscence, and the biogenetic law. J Hist Bio 11(1):129–158

Mao Z (1964) Talk on Sakata's article. In: Selected works of Mao Tse-tung, vol IX. http://www.marxists.org/reference/archive/mao/selected-works/volume-9/mswv9_28.html. Accessed 30 Jan 2014

Nature (2012) John Maddox Prize: two strong-minded individuals are the first winners of an award for standing up for science. Nature 491:160, 8 November

People's Daily (2009) Founder of China's biophysics and academician of Chinese academy of science Bei Shizhang passed away (我国生物物理学奠基人中国科学院院士贝时璋逝世). People's Daily, 1 November

Potthast T, Hoßfeld U (2010) Vererbungs- und Entwicklungslehren in Zoologie, Botanik and Rassenkunde/Rassenbiologie: Zentrale Forschungsfelder der Biologie an der Universität Tübingen im Nationalsozialismus. In: Wiesing U et al (eds) Die Universität Tübingen im Nationalsozialismus. Franz Steiner Verlag, Stuttgart, pp 435–482

Rao Y (饶议) (2007) In memory of Zou Chenglu. Commentary Sci Cult 4(2):38–45

Rao Y (饶议) (2009) Rao discussed science (饶议科学). Shanghai Science and Technology Education Press, Shanghai

Rapoport Y (1991) The doctor's plot of 1953: a survivor's memoir of Stalin's last act of terror, against jews and science. Harvard University Press, Cambridge

Science Times (2009) Founder of Chinese biophysics Bei Shizhang died at 107 (中国生物物理学奠基人贝时璋逝世 享年107岁). Science Times, 30 October

Science Times (2010) Bei Shizhang: researching life science with his life (贝时璋:用自己的生命研究生命). Science Times, 11 November

The New Threads (2005) http://www.xys.org/forum/db/208/210.html, 9 November. Accessed 10 April 2015

Wang J (王静) (2009) Remembering academician Bei Shizhang: pioneering the field of biology, expressing the passion of science (记贝时璋院士:创生物伟业 抒科学豪情). Science Times, 2 November

Wang GY (王谷岩) (2010a) Bei Shizhang: a biography (贝时璋传). Science Press, Beijing

Wang GY (王谷岩) (2010b) Shizhang Bei (Sitsan Pai) and his theory of cell reformation. Protein Cell 1(4):315–318

Xinhua News (2009) Thousands farewell Bei Shizhang (千人送别贝时璋). Xinhua News, 5 November

Young D (1911) The implantation of the glochidium on the fish. The Univ Mo Bull: Sci Ser 1(1):1–20

Ying YM (应幼梅) (1992) The life, work and thought of Professor Bei Shizhang (贝时璋教授的生活、工作和思想). In: SWB, pp 1–39

Zhang QQ (张青棋) (1996) The history and reality of research in the philosophy of biology in China after 1949 (建国以来我国生物哲学研究的历史与现状). Studies in Dialectics of Nature. http://hps.pku.edu.cn/2003/06/616. Accessed 21 Dec 2014

Zhang JJ (张坚军) (2002) A biography of Bei Shizhang (贝时璋传). Ningbo Press, Hangzhou

Zhao YH (赵亚辉), Chen XX (陈星星) (2009) The oldest academician Bei Shizhang worked to the last moment: 'We Have to Fight For Our Country.' (最高龄院士贝时璋打拼到最后一刻: '我们要为国家争口气'). People's Daily, 2 November

Zhinkin LN, Mikhailov VP (1958) On 'The New Cell Theory': two soviet authors critically review recent soviet work on the origin of the cell. Science 128(3317):182–186

Chapter 3
The Institutional Infrastructure of Biophysics

Abstract This chapter illustrates the institutional infrastructure of biophysics in terms of research establishment, education system, specialized journals, and professional societies. The reason to consider the institution underpinning of biophysics is that: institutional shaping is a critical factor in the development of a scientific discipline; institutions justify and finance biophysical inquiry as the disciplinary history of biophysics in the United States and Germany showed. Since institutional blessing is an essential condition for the robust development of biophysics, this chapter examines how the research, education, journalistic, and professional aspects of biophysics paint a picture of the disciplinary landscape of biophysics in post-1949 China. From the establishment of the Institute of Biophysics at the Chinese Academy of Sciences, the founding of a biophysics department at the University of Science and Technology of China, the launching of a biophysics journal entitled *Progress in Biochemistry and Biophysics*, to the inauguration of the Biophysical Society of China, these concrete steps are part of the larger plan to make biophysics into a scientific discipline in contemporary China.

Keywords Chinese Academy of Sciences · First U.S.-China biophysics contact · Institute of Biophysics · *Progress in Biochemistry and Biophysics* · The Biophysical Society of China

The last chapter introduced the founding figure of biophysics in China and his disputed academic achievement. The birth of Bei's theory of cell reformation took place before the Communist takeover of China. Cell reformation was not a communist concoction, nor was it a critical biophysical invention. Bei's unorthodox knowledge was predicated on a particular theoretical assumption rather than advanced microscopy. Cell reformation and Bei's developmental studies only became "biophysical" when Bei established an institute of biophysics after 1949. Bei's early life and intellectual background mattered as these factors nurtured the life of Bei and therefore sowed the "human" seed for the growth of the biophysical plant.

The presence of powerful elites set the stage for the larger effort that was to bear fruit later on. Having capable individuals is only one of the many prerequisites for disciplinary development. How does the supply of specialized personnel transform

© The Author(s) 2015
C.Y.L. Luk, *A History of Biophysics in Contemporary China*,
SpringerBriefs in History of Science and Technology,
DOI 10.1007/978-3-319-18093-9_3

a scientific discipline? In the case of biophysics, institutional shaping is a critical factor; institutions justify and finance scientific inquiry. Institutional blessing is an essential condition for robust scientific development.

This chapter explores the disciplinary landscape of biophysics in post-1949 China. Drawing on institutional yearbooks, memoirs, collections of diaries, and technical journals, my attention is placed on the research establishment, educational system, specialized journals, and professional society that comprise the disciplinary structure of biophysics in the PRC.

3.1 The Research Establishment of Biophysics in the PRC

The research establishment of biophysics reflects the strong institutional position of leading biophysicists. The political reputation of Bei Shizhang, the founder of biophysics in China, is exemplary. In 1948, Bei was among the first cohort of scientists to be elected to academicians at *Academia Sinica*. With this honor and the prestige associated with it, Bei was invited to partake in The Preparatory Committee of the National Congress of Workers in Natural Sciences which was held in Beijing on 13 July 1949. Two hundred and eighty five selected representatives from various political and professional sectors attended this convention (Wang 2010). Slightly predating the official inauguration of the PRC, the idea of creating a national academy of science was raised on this occasion. Bei was one of the conveners of the science panel, and he along with several panel members such as Yan Jici and Zhu Kezhen proposed that the government should establish a national academy of science. The panel then had an after-dinner session with vice-chairman Zhou Enlai to discuss plans for setting up a Chinese Academy of Sciences (Fan 1999).

The episode of Bei's participation in the establishment of the Academy was an important stepping stone for the institutionalization of biophysics. It is important to identify Bei's political leverage in the scientific clique surrounding the Academy. Before biophysics was formally instituted, the father of biophysics was already an active participant in the planning of the Academy. Bei's political sway within the group of Academy bureaucrats led him to more concrete opportunities for establishing a research institute dedicated to biophysics.

Soon after the inception of the Academy of Sciences, Bei was one of the twenty-six scientists' delegation visiting Moscow in 1953 under the policy of "learning from the big brother." This Moscow trip gave him the rare opportunity to meet with the Soviet cytologist O.B. Lepeshinskaya.

As I revealed in the last chapter, Bei had been exposed to the existence and importance of Lepeshinskaya and her revolutionary claim of "vital substances" before the Communist takeover of China. Bei's enthusiasm in expecting Lepeshinskaya was guileless. In his first meeting with Lepeshinskaya, Bei asked her about the exact chemical components and methods of extraction of the so-called "vital substances." Lepeshinskaya was reported to have sidestepped these inconvenient questions.

The second time Bei met with Lepeshinskaya, she did not want to discuss any aspect of her work on the formation of blood islands from yolk spheres in chicken embryos. Instead, she wanted to broadcast her latest discovery on the rejuvenating power and crop-yielding potentials of soda baths (Rapoport 1991, p. 260). Much to Bei's chagrin, Lepeshinskaya was not willing to share the details of her work or to clarify the confusions in her "revolutionary" cell study.

Lepeshinskaya was known as a rather arrogant scientist. Other Chinese scientists also reflected on the patronizing attitudes of Lepeshinskaya and Lysenko during the Moscow visit (Zhang and Deng 1996). Bei sensed her condescending attitude, but he was relentless to know more about her alleged "ground-breaking" study—a claim that was lauded by Stalin and Lysenko as comprising "a socialist revolution in cytology." Yet Bei's eagerness to converse with his respectable Soviet colleague was unreciprocated. Lepeshinskaya wanted to dominate the dialogue unilaterally without expressing a tinge of interest in Bei's cellular studies.

After the encounter with O.B. Lepeshinskaya, Bei turned his attention to the biophysics establishments in the Soviet Union. In 1957, the PRC government sent another delegation team to Moscow to discuss issues of cooperation in science and technology with the Soviet Union. Bei was a leading representative of the biological science division, and between 24 October and 13 November in 1957, he led a discussion with Soviet experts regarding the project on "the foundation of building biophysics." (Wang 2010, p. 139). In the same panel, Bei authorized four new areas in biology, namely microbiology, biophysics, biochemistry, and genetics (Ying 1992, p. 35). This was one of Bei's earliest moves for making "biophysics" known in the political sphere and was a preparatory step for laying the groundwork of introducing the discipline known as biophysics to China.

Besides theoretical deliberation, he also seized the chance to visit the research facilities in the Soviet Academy of Sciences and higher-education institutions in the Moscow area. Of particular interest to him were the research and teaching facilities in radiobiology, biophysics, , and medicine. As Wang Guyan suggested, "the (Moscow) visitation prepared for Bei's effort in pioneering a new discipline and establishing the Institute of Biophysics and the Department of Biophysics at the University of Science and Technology of China in 1958." (Wang 2010, p. 156)

Slightly after the first Moscow trip, Bei had taken actions to legitimize biophysics in China by incorporating biophysics into the institutional structure of the Academy of sciences. In 1955, the central government promulgated a twelve-year development blueprint known as "A Draft Developmental Outline of National Agriculture between 1956 and 1967." This policy document was said to inspire other bureaus to release similar twelve-year plans for their respective sectors (Wang 2010, p. 152). In 1956, the Academy's secretariat office prepared the "Key Scientific Research Projects for the Chinese Academy of Sciences to Undertake in Twelve Years (for Natural Science & Technical Science)." As the designated academic secretariat of the Academy, Bei was a core member in the drafting and consultation of the twelve-year plan. His role in the planning of this policy document was exemplified in his chairing of the session on "Several Important Questions in Basic Theories of Natural Sciences" in April 1956. Another co-chair of this session was the celebrated physicist Zhou Peiyuan. Bei was also invited to sit on the committee that was responsible for finalizing the resultant policy document—"A Long-Term Outline of Planning for the Development of Science and Technology Between 1956 and 1967." After this policy document was disseminated, Bei was appointed as the head of the division of biological sciences in the planning committees of the State Council.

Where did biophysics feature in this policy domain? In the policy document entitled "The Second Five-Year Planning Outline for the Chinese Academy of Sciences (A Draft)" released in March 1957, biophysics was highlighted as a key marginal discipline awaiting further development (Wang 2010, p. 248). This was a window of opportunity for Bei. Serving as the head of the biological planning committees of the State Council, he proposed to the State Council that the Academy should establish a research institute for biophysics, which the State Council then immediately approved. In the ninth executive meeting of the Academy, the executive committee licensed the plan for constructing the Institute of Biophysics, and on 20 September 1958 the Institute for Biophysics at the Chinese Academy of Sciences was formally established. China was one of the few places in the world where biophysics was legitimized and endorsed at both the national and institutional levels.

What is noteworthy in the above description is the way in which the institutional position of Bei Shizhang in the Academy gave him the opportunity to create a research institute by the name of biophysics.

3.2 The Educational System of Biophysics

The capacity to achieve discipline building appears to be dependent upon the institutional leverage of the leading scientists. The educational system of biophysics reflected the same political and social forces at play. It was mostly political cachet that allowed Bei Shizhang to carve out an institutional space for biophysics under the Academy of Sciences. Although there was a new institute and new projects lined up for the self-identified "biophysicists," scientific recognition of biophysics as a distinct discipline did not come about immediately. Among the more learned biologists in socialist China, it was still the case that few of them knew what "biophysics" was, and many did not even make an effort to know the field. The indifference and even opposition to biophysics was best summarized by an acerbic voice at the time: "There was only physiology, no biophysics." (只有生理学, 没有生物物理学). Not only did physiology have a longer tradition in modern China, it had rallied to itself a better academic understanding and reputation in biological circles.[1] Biophysics, on the other hand, was relatively unknown to biologists and physicists. Bei Shizhang realized that the mere adoption of the term "biophysics" was not enough to promote its academic profile: He saw educational campaigns as the key to raise awareness of biophysics as an independent discipline and to attract able and ambitious students to this newfound profession.

In the same year as the Institute of Biophysics was founded, a university department with specialized study programs in biophysics was created at the University of Science and Technology of China (中国科学技术大学) from 1958 to 1962. One could see a convergence of institutional impetus and intellectual ambitions in establishing a biophysics department at USTC. One of the notable features of USTC is that it is not just another higher-education institution in China with a specialization in science and technology. Most universities, including the prestigious Peking and Tsinghua Universities, are under the administration of the Ministry of Education. By contrast, USTC—from its establishment until this very day—falls under the jurisdiction of the Chinese Academy of Sciences.[2] The organizational nature of the Chinese Academy of Sciences does not resemble that of the National Academy of Science in the US. Rather, it adopts the institutional model of France and Russia in which the national academy is not just an honorific society but also a major headquarters for research and teaching activities. In China, the Academy was known as the "head of the train" of national R&D activities when it was founded in 1949, but it wasn't clear whether the Academy should assume leadership in scientific and technical education. In 1950, a controversy erupted between Guo Moruo, then president of the Academy, and Yang Xiufeng, then secretary of Higher Education at the Ministry of Education, regarding whether the Academy or the Ministry should act as the center of scientific education and training. Mao stepped in and asked both parties to set aside their jurisdictional interests and make collective contributions to the common cause of national modernization (Zhang and Deng 1996, pp. 83–84). Although Mao brought about an immediate resolution to this particular conflict, the undertones of the quarrel remained, and it would foreshadow the entanglement of the Academy in educational affairs in 1956.

In 1956, the policy document entitled "Twelve-Year Long Term Plan for the Development of Science and Technology, 1956–1967" was announced. The policy highlighted higher education in natural science and advanced technology as one of the *sine qua non* for socialist construction. The legislative imperative created an

---

[1] One of the indicators of the leading disciplinary status of physiology in China is the early history of its professional journal, *The Chinese Journal of Physiology* (中国生理学杂志), which was inaugurated in 1927 by four distinguished physiologists and medical biochemists under the auspices of the Rockefeller Foundation. The journal was renowned for publishing research papers of exceptionally high quality in Republican China.

[2] As of 2013, seventy-two universities and higher education institutions are under the jurisdiction of MOE while only three are under CAS. Besides USTC, the other two CAS-managed universities are the University of the Chinese Academy of Sciences (中国科学院大学) and the Shanghai University of Science and Technology (上海科技大学).

instant demand for advanced training in science and technology (Xu 1982). Reformists and board members at the Academy felt that given the pivotal role of the Academy in carrying out the "Twelve-Year Plan," it should take the initiative to improve college education in science and technology. Leaders of the Academy saw their organization as an agenda-setting body in responding to the pressures for science instruction in order to meet the goals of national modernization. The motion was first proposed by Guo Moruo in the third executive meeting of the Academy on 18 March 1958, where he raised the possibility that the Chinese Academy of Sciences could consider developing an affiliated educational institution to facilitate the training of cadres. Guo's idea was supported by Nie Rongzhen with the endorsement of Zhou Enlai. On 21 May 1958, Nie submitted a formal proposal to then Secretary General of the Politburo, Deng Xiaoping, who passed the motion of constructing a Academy-directed university of science and technology (Zhu 2008).[3]

The driving force behind the Academy's desire to secure a role in training students in science and engineering was the requisite human resources for implementing the "Twelve-Year Plan." The immediate concern of the Academy leaders and scientists was the need to prepare a cohort of scientists and engineers who could meet the practical requirements of the many modernization projects articulated in the "Twelve-Year Plan." However, there were ulterior motives: Members of academic divisions in the Academy also recognized that educational programs in advanced science and technology could advertise the devotion of scientists to public service. Publicity meant better reputation and wider socio-political influence of scientists. Serving advanced science education could further highlight the political reliability of scientists as an elite class by "encouraging people to orient toward the Academy." (Zhang 2009)

It is evident that the founding of USTC was inseparable from the institutional mission of the Academy, as not only was USTC under the leadership of the Academy from the very beginning, it was established primarily as an instrument to fulfill the service role envisioned by leaders of the Academy.

In fact, the research activity was intimately linked to the educational commitment in such a way that both the research and education of biophysics were intended to "serve definite purposes and lead to the solution of important problems of a mass characters," as uttered by the vice-president of the Academy (Chan 1992, p. 47). Federal projects for socialist modernization incentivized the proliferation of systematic education and apprenticeship by offering a utilitarian thrust to fortify the enfeebled educational base in science and technology.

Nevertheless, the political economy of science education does not capture the full range of partnership between CAS and USTC. The genesis of USTC illustrated how the principle of "taking missions as the longitude, disciplines as the latitude, and using missions to drive disciplines" (以任务为经, 以学科为纬, 以任务带学科) was put into practice (Fan 1999). A critical motif in the mission-drive-discipline

---

[3]For an abridged history, see "Founding background and preparation process," at http://arch.ustc.edu.cn/history.htm.

principle is the interweaving of missions and disciplines. The symbiosis between USTC and CAS is crystallized in the policy motto of "running the university with the resources of the entire Academy and integrating departments with relevant research institutes" (全院办校,所系结合). Partly drawing from the Soviet experience of merging the Novosibirsk State University with the Siberian Division of the USSR Academy of Sciences, the objective was to furnish an integrative program of science education by marshaling human and instrumental resources from all over the Academy (Wang 2010). In this way, each academic department of USTC would correspond impeccably to the same specialty of the respective institutes at CAS. The CAS-USTC interplay closed the gaps between teaching and research, theory and practice, basic and applied sciences, academic ideals and civil service. Thirteen major research institutes of CAS begat a department at USTC with the director assuming the role of department chair. Following this policy of dual superintendence, Guo Muoruo, then president of CAS assumed the presidency of USTC while thirteen directors of the research institutes were appointed as the respective chairs of departments at USTC. One of these founding departments was biophysics and Bei Shizhang was the department chair.

As one of the department chairs at USTC, Bei Shizhang was eager to introduce an educational framework for biophysics. In the late fifties and early sixties, the founding of biophysics as a department was a strategic move to stabilize and strengthen the disciplinary status of biophysics. By offering an institutional site for training a new cadre of biophysicists, the school drew scores of young Chinese eager to learn a style of biophysics that was broadly biological and with a service role in the research missions undertaken by the Institute of Biophysics at the Academy. It served Bei's aim as a discipline builder by allowing him to define a style and a territory for the discipline that he advocated. Biophysics was taught at USTC from the fall of 1958 to the eve of Cultural Revolution in 1966. For more about the teaching and learning of biophysics at USTC, see Luk (2015).

As a result of the political leverage of Bei Shizhang, new research and educational institutions for biophysics were created at CAS and USTC respectively. The enabling condition for these disciplinary-building activities also led to the insulation of biophysicists and biophysics against party assertiveness in the midst of the Cultural Revolution.

During the Cultural Revolution, science and technology in relation to national defense and national prestige were given special protection against revolutionary interruption by the People's Liberation Army (PLA), but the defense ministry was not a sanctuary for all scientists and all lines of scientific projects under its umbrella. Scholarly debates erupted regarding the Janus-faced nature of military research and missions during the Cultural Revolution. Darryl Brock observed "there is a broad sense among scholars that defense scientists enjoyed protection from the Cultural Revolution due to the national security considerations related to their work," (Brock 2013, p. 82) but in the same academic volume, Cong Cao quibbled, "those members of the scientific community who worked on the military research projects were spared of the worst of the turmoil, but they were not completely safe. The Cultural Revolution soon spread to the military as well, with research, development, and production coming to a standstill because of Red Guards' rampages." (Cao 2013, p. 127)

While pitched battles between various groups of Red Guards and PLA have been explored at some lengths in the literature, struggles between scientists and military officers working in the Defense Ministry were seldom the subjects of analysis (Esherick et al. 2006). We know that the PLA were called in to restrain the Red rebels and to restore order in universities and high schools but we are not well informed about the role of scientists and their interaction with the PLA officers. The PLA-civilian interaction is one thing (Gao 1987); but for the purpose of this study, what piqued my interest is the extent of the military-science interplay. Did defense scientists enjoy de facto protection? How effective was the sheltering? How did biophysicists fit into this military-science regimen?

One common point of reference for reporting the PLA-science misalignment is the misfortunes at the Seventh Ministry of Machine Building (Seventh MMB). The Seventh MMB was one of the worst hit agencies in the defense sector during the Cultural Revolution. MacFarquhar and Schoenhals in their study of the elite politics of the Cultural Revolution noted, "the Seventh Ministry of Machine Building, responsible for missile and satellite development, was also riddled with problems. Disruptive factional infighting had been endemic since the very beginning of the Cultural Revolution" (MacFarquhar and Schoenhals 2006, p. 385). A Chinese source on the history of space science and technology in China also stated, "the scientific research of the Seventh MMB has been interfered with and damaged immensely by the 'Cultural Revolution'." (Li 2005, p. 343). Even a celebratory volume on the heroic achievement of nuclear and missile scientists and engineers suggested that "factionalism of the Cultural Revolution wreaked havoc on the Seventh MMB." (Song 2001, p. 405). Yet the devastation of the Seventh MMB is typically only mentioned but not examined. A recent scholarly work offered a rationale for the paucity of analysis of defense-related aspects of the Cultural Revolution: "the classified nature of defense science rendered documentation relatively unavailable to reporters… and modern historians. It is thus not surprising that relatively little analysis has been conducted related to S&T of national defense related to the Cultural Revolution." (Brock 2013, p. 83)

Indeed, the literature gap is largely due to a lack of access to sources and documentation that were not open to the public until recently. Scholars in China Studies have noted that what distinguishes the new wave of Cultural Revolution scholarship from the first generation of research is precisely the steady increase in the availability of once-scarce official and unofficial sources of information (Esherick et al. 2006, p. 6).

Historians now enjoy the luxury of the voluminous primary and secondary materials that were not available to the previous generation. The exponential growth of documentation allows for a more concrete and substantial portrayal of aspects that were obscure in the first wave of research. What happened at the Seventh MMB between 1966 and 1969, perhaps, is a case in which newly released evidence could enable a more elaborate analysis beyond the line of interpretation established by those who worked with more limited sources.

In light of the widely noted but thinly examined mishaps at the Seventh MMB, one could benefit from the personal accounts of key participants in the struggles on the spot. The diary of Yang Guoyu (杨国宇) chronicled some of the conflicts and confrontation between PLA officers and various cliques based on his firsthand knowledge and experiences at the Seventh MMB (Yang 2000). Although bio-physicists were not depicted in Yang's recollections nor were biophysicists directly involved in the armed struggles at the Seventh MMB, Yang's diary commands attention as the circumstances at the Seventh MMB reflected the overall happenings of defense science—of which biophysics is a part—during the Cultural Revolution.

All personal accounts and statements of oral histories should be treated with circumspection and skepticism. The methodological pitfall of relying on this type of literary evidence is that sometimes the author masquerades fictional accounts as ostensible facts by the virtue of his or her authorship. As Stephen Pyne has cau-tioned, a memoirist's offense is where "the author claims the authority and gravitas of nonfiction, stating that what he or she says is true, but in reality fictionalizes the story or in some cases invents it out of whole cloth." (Pyne 2011, p. 36). Nonetheless, personal recollections afford an opportunity to assess the Cultural Revolution as experienced at an individual level. In what follows, I do not take Yang's depiction at face value and as fully credible; rather, I inspect its authenticity by comparing his account with other testimonies. I believe these potentially flawed, yet integral materials, if handled carefully, could further illuminate the organiza-tional dynamics of defense science and technology in revolutionary China.

A long-time lieutenant of Zhou Enlai and a Long March survivor, Yang oversaw the military takeover of the Seventh MMB during the period from March 1967 to November 1969 (Lewis and Xue 1994, pp. 148-151). On 22 March 1967, Yang received a direct order from the NDSTC authorizing the twenty-third military base of the PLA to preside over the Third Research Institute of the Seventh MMB. Yang and his military unit were specifically chosen to take control of the Seventh MMB—the bureau responsible for the research and development of missiles and spacecraft—due to his earlier participation in the missile and space program. Yang reported his chairing of the initial meeting between the two major factions dated 19 May 1967. Over ten thousand protestors attended the meeting and expressed their revolutionary energy by constantly interrupting Yang's speech and chal-lenging the authority of the military control committee. Yang reacted by reiterating the orientation of the general policy of the military control committee as serving public interest and aligning with Mao Zedong's thought. After an overnight sit-in, the mobs began to disperse after they realized that the military control officers were not going to budge. Although there were sporadic disputes and verbal attacks, bloody feuding was kept under control according to Yang: "we are here to contain some events because we came from the outside, so they thought we were simple-minded and had no interaction with capitalist-roaders or minority leaders within the military." (Yang 2000, p. 2). The strategy Yang utilized to control the mass emotion was three-fold. First, the military control committee would not take sides in any matter despite being charged by the mutinous crowd. Second, the military control committee adopted a firm stance by expressing its unswerving loyalty to the thought and leadership of Chairman Mao when faced with opposition and dis-obedience. Third, the military control committee would allow criticism through Big Character Posters or other written channels but not in the form of physical assaults. When bitter fighting paralyzed the Second Research Institute of the Seventh MMB on 30 August 1967, Yang tried every means to prevent the bellicose flame from reaching the Third Institute. Occasional armed struggles notwithstanding, the pre-vention of large-scale violence and bloodshed at the Third Institute of the Seventh MMB reflects the partial success of the tripartite strategy Yang had devised.

But the personnel turnover of the military control committee quickly altered the course of events. The existing military control committee was replaced by a new committee on 4 April 1968. Yang was seconded to the post of assistant commander but was still sitting on the new military control committee. He ended up in a quarrel with the young officers in the new committee over the new practice of singing selected quotations of Chairman Mao before breakfast, among other issues. Refusing to join the mandatory quotation-reciting sessions each morning, Yang lost his sway in the new military control committee, and on 16 May, Yang's secretary Jiang Changying (蒋昌应) was badly beaten as he tried to stop a brawl between the notorious "915" and "916" gangs. This was an alarming sign of the escalating interfactional antagonism, which would result in tragedy on 8 June 1968, when Yang was informed of the death of Yao Tongbin (姚桐斌), the Birmingham University-trained metallurgist and director of the Third Institute of the Seventh MMB, as a result of a severe beating by some of the "915" members (Yang 2000, pp. 7–8).

Immediately after Yao's death, Yang was busy making logistic arrangements for transporting Yao's body. Apparently, many state-run hospitals including the Beijing Municipal Hospital and the Air Force General Hospitals were reluctant to collect the corpse for fear of an attack by the mob at the Seventh MMB. Ultimately, Yang had to force the Navy Hospital to accept the cadaver by means of military order. Meanwhile, Yang had to seek consent from Yao's wife in order to proceed with an autopsy of the body. Yang's diary depicted initial resistance, but eventual compliance to his request by Peng Jieqing (彭洁清), the widow of Yao Tongbin who was grieving over the loss of her husband, father to her three young daughters. Three decades later, Peng Jieqing published a memoir expressing her love and devotion to her husband (Peng 2002). Peng's reminiscence of the events after the killing of Yao mentioned her encounter with Yang Guoyu for negotiating Yao's postmortem arrangement. Madam Yao characterized Yang as a middle-aged, round-faced military officer dressed in Navy uniform who spoke with a Sichuan accent. Peng was opposed to the idea of cutting open her husband's body at first but finally acknowledged the importance of performing a proper medical investigation in order to dispel rumors of Yao committing suicide or dying of a heart attack. As the medical team carried Yao's body away, Peng composed and distributed a Big Character Poster illustrating Yao's life and the gut-wrenching story of his death. She was later told that many people burst into tears as they read her poster (Peng 2002, pp. 24–27).

Besides lending support to Yang's documentation, Peng's recollection also casts light on hidden aspects of the Cultural Revolution and even contemporary Chinese history. David Apter and Tony Saich uphold the view that Chinese widows' oral history can open a new avenue for revisiting modern Chinese history: "A rich, and indeed large, resource for uncovering actual events rather than myths are the many widows. China is a country of angry widows. Each shift in the party line produced its own legacy of such widows whose husbands were pulled down, humiliated, subjected to trials, committed suicide or died of beatings, or were atrophied by longprison sentences." (Apter and Saich 1994, p. 20)

Peng was among the ranks of Chinese widows who lost their husbands to various kinds of unfair sanctions in the twentieth century. Through relating mundane aspects of everyday life, Peng's memoirs, like Anne Frank's diary, revealed the horror and devastation an overzealous mob could inflict upon an ordinary soul. But her writing was also interspersed with discussion of the many patriotic deeds of Yao and his enthusiasm for the future of China's aerospace industry, of which her deceased husband was a crucial member. On 2 December 2005, Peng gave an invited lecture at a middle school in her husband's hometown. When asked whether the victimization of her husband caused her any disillusionment in the party-state, she answered with an absolute "no" and said it was a few malicious people, rather than the state, that were responsible for her husband's death (Xinhua News 2005).

Although some may cast doubt on the veracity and objectivity of Peng's publicly captured remark, I suspect that part of the reasons for her persistent faith in the state has to do with the ways in which the authorities handled Yao's case. When the news of Yao's death reached Zhou Enlai, Zhou was reportedly shocked to his very core. He dropped his teacup on the floor and exhibited indignation about the loss of a brilliant scientist—all the more so because Yao was personally appointed by Zhou to work in missile research in 1958 (Peng 2002, p. 146). The death of Yao impelled the Premier to issue a special protection plan for a small group of scientists and engineers in cutting-edge science and technology projects (Zhang 1986, p. 65). Decades after the Cultural Revolution, an unnamed veteran space engineer applauded the effort of Zhou Enlai this way, "I suffered quite insignificantly because a colleague of mine sacrificed his own life, and a widely-respected leader took action to protect the scientific community." The former referred to Yao Tongbin and the latter to Zhou Enlai (Peng 2002, p. 185). Song Jian, a notable cyberneticist and a high-profile military scientist, elaborated on the positive consequence of Yao's sacrifice: "Yao died for a noble cause, because his unfortunate death prompted Premier Zhou to enforce the protection of many other scientists and ensured their safety." (Song 2001, p. 88). This point was echoed by Stacey Solomone recently: "Without Yao's sacrifice, Premier Zhou may not have extended protection to fellow aerospace scientists and engineers." (Solomone 2013, p. 243). In the spring of 1976, Peng expressed her deepest condolences at Zhou's passing and praised Zhou's attention to her and the family since Yao's murder (Peng 2002, p. 186). Recently in China, Yao has been honored with a posthumously awarded medal for his contribution to the bomb and missile program, along with a museum with the name "Former Residence of Yao Tongbin" in his birthplace in Zhejiang (Song 2005).

Although unofficial sources of information like Yang's diary or Peng's memoir are susceptible to personal caprices and vagaries of individual memories, it is possible to test the accuracy of these personal recollections from documentation in other references. Peng Jieqing's memoir authenticates some of Yang's statements. More importantly, John Lewis and Xue Litai identified Yang as the chief officer in charge of the military control committee of the Seventh MMB by order of Zhou Enlai in their study of China's sea-based strategic programs (Lewis and Xue 1994).

Translating what is singular and arbitrary to a collective and more plausible account, these new sources of material could fill a historiographical gap in the science—military intersection during the Cultural Revolution. As more relevant writings are released from the military sector in the future, one could even hope to elucidate more unknown aspects of the Chinese strategic missile and space program (Esherick et al. 2006).

The saga of Yao Tongbin and the untold (at least in English literature) story of Yang Guoyu are related because biophysics was at the forefront of the science-military interface between 1967 and 1970.

In March 1967, Nie Rongzhen, in his capacity as the vice premier and former head of the now defunct State Science and Technology Commission (SSTC) and National Defense Scientific and Technology Commission (NDSTC), proposed institutional reforms in the science and technology operations at the civilian–military intersection. Since many defense research projects relied on the cooperation between the defense industries and the civilian research institutions—the CAS in particular—it became a priority for Marshal Nie to issue a special plan for the sustained development of military science at times of political perturbation (Nie 1988). The proposal, entitled "On the Institutional Adjustment and Restructuring Schemes of the National Defense Ministry," was introduced to reduce the level of disruption of scientific research in the defense establishment. Under the section heading of "Biomedical Aspects of the Manned Spaceflight Program" in this document, Nie proposed setting up a "Cosmomedical and Cosmobiological Research Center" in the defense establishment by transferring the cosmobiology unit of IBP from the CAS framework to the NDSTC infrastructure (IBP-CAS 2008, p. 15). Chairman Mao endorsed Nie's plan on 25 October 1967 (*Science Times* 1999, p. 120). In April 1968, around one hundred researchers and technicians from three research centers in IBP—The Cosmobiology Center, the Space Animal Testing Center, and the Cosmobiological General Control Center—along with all the equipment and libraries, were handed over to the Space Medical Engineering Research Institute (SMERI) under the administration of NDSTC (IBP-CAS 2008, p. 221; Li 2005, p. 794). Also recruited into the SMERI were operative units from the Academy of Military Medical Sciences and the Chinese Academy of Medical Sciences—partnering the units with IBP in Mission "581." The cosmobiologists have remained in the military-industrial complex ever since. Between 1992 and 2012, the former biophysicists in the SMERI were responsible for the selection and training of the first cohort of taikonauts in China's first comprehensive manned spaceflight program (known as project "921"). Several biophysicists assumed high-ranking posts in China's manned space program: the principal designer-in-chief and the second chief commander of the "yuhangyuan system," the chief designer of the 1/3 "subsidiary system," and the chief designer of the "environmental control and life support subsidiary system" in the "spaceship system." (Wang 2010, pp. 263–265)

The above cosmobiological units that were handed over to NDSTC during the Cultural Revolution only accounted for around one third of the manpower of IBP-CAS. A substantial group of biophysicists stayed with the Academy throughout the Cultural Revolution. The Institute of Biophysics was one of the ten institutes that remained under the Academy's jurisdiction in 1972. The other nine CAS-administered institutes at this time were the Institute of Mathematics, the Institute of Physics, the Institute of Chemistry, the Beijing Observatory, the Institute of Microbiology, the Institute of Genetics, the Institute of Geology, the Institute of Atmospheric Physics, and the Institute of Vertebrate Paleontology and Paleoanthropology (Fan 1999, p. 212). To put it another way, one arm of IBP was transplanted to the military-industrial complex while the main body of IBP was still in the civilian system throughout the entire period when civilian science was discredited by proletarian distrust. This is not a small matter if one considers the scale of reduction of the Academy (Miller 1996, pp. 88-89). Many important institutes, like the Institute of Mechanics, were ordered to separate from the Academy, but biophysics managed to stick with CAS during the stormy years.

David Lampton characterized the Cultural Revolution as "a massive movement in which what was at stake varied according to one's organizational affiliation, values, and interests. Consequently, the campaign defies facile description." (Lampton 1977, p. 202). The Cultural Revolution was by and large a disastrous event for scientists and other learned professionals due to the abrasive nature of the proletariat's distrust of professional expertise. But it was not necessarily true that all Chinese scientists were victimized by the campaign, because as Lampton observes, one's power to resist revolutionary fervor resides in one's institutional position, and particularly at politically unstable time. The biophysical infrastructure survived throughout the Cultural Revolution. The following section documents the role of Bei Shizhang in initiating the first US-China biophysics contact during the latter half of the Cultural Revolution. Bei's involvement in this historic event is taken as a part of the larger movement to professionalize biophysics that entails the launching of a specialized journal and a professional society.

3.3 Bei Shizhang and the First US-China Biophysics Contact

Among the many plausible reasons that might have contributed to the lenient treatment of biophysicists during the Cultural Revolution, the political standing of the leading scientists seems to be a critical factor. If the political credibility of a scientific leader was undermined, it could cripple the institute under his or her directorship. This is the infamous practice of "guilt by association" by incriminating people who are in alleged affiliation with the person in question. The inverse side of this disreputable formula is that those affiliated with a scientist in good political standing were usually spared the political attacks occurring elsewhere. In the case of biophysics, it is worth asking: what was Bei's political trustworthiness during the Cultural Revolution?

Bei's biographer and follower Wang portrayed Bei's experience during the transitional years of the Cultural Revolution as "very lucky. Not only was he not subject to criticisms, there were many people protecting him." (Wang 2010, p. 270). Wang's source of evidence drew from Bei's appointment as the leader of the first scientific exchange orchestrated by the Mao-Nixon diplomacy. In 1972, Bei led a seven-scientist delegation to visit the US following the table-tennis tournament initiated in 1971 (*People's Daily* 1972). The visit marked the ushering in of a new era of development and exchange of scientific and technological affairs between China and countries in the US sphere of influence. The 1972 delegation was the first science emissary sent by China to visit the US and its allies since the onslaught of the Cultural Revolution. Between 6 October and 17 December, the delegation toured university laboratories and research facilities in areas ranging from high-energy physics, plasma physics, controlled thermonuclear reaction, computer science, environmental science, to biophysics in the UK, the US, Canada and Sweden (Fan 1999, p. 271). In the US, the Chinese science delegation was cordially received by the Federation of American Scientists and the Committee on Scholarly Communication with the People's Republic of China (CSCPRC)—a committee jointly founded by the National Academy of Sciences (NAS), the American Council of Learned Societies (ACLS), and the Social Science Research Council (SSRC) in response to the Shanghai Communiqué issued on 27 February 1972 (CSCPRC 1986).

Besides acting as the titular head of the Chinese delegation, Bei Shizhang's visit to the US in the fall of 1972 bore disciplinary significance to the first trans-Pacific exchange between indigenous Chinese and Chinese-American biophysicists. Hsin-Ti Tien (田心棣), chair of the department of biophysics at Michigan State University, wrote to Bei Shizhang in early 1973 expressing his interests to visit IBP-CAS and other biophysics-related establishments in China after Tien learned of Bei's ambassadorial role in the first Chinese science delegation. Tien departed for China in 1973 as an official representative of the Biophysical Society upon the invitation of Bei Shizhang and with the endorsement of the US Biophysical Society Council (Tien 1975).

Under the auspices of the US Biophysical Society, Tien visited several major biophysics-related research institutes and university departments. In addition to the Institute of Biophysics and the Institute of Plant Research at Beijing, Tien also traveled to the Institute of Plant Physiology at Shanghai, Sun Yat-sen University at Guangzhou, Wuhan University, Central China Normal University at Wuhan, and Peking University. In his field report submitted to the Biophysical Society Council, Tien considered the effects of the Cultural Revolution on scientific research particularly pertaining to the Institute of Biophysics and the Academy after giving a general profile on the basic history and circumstances of each research institute he visited.[4] Tien reached the conclusion that "During the Cultural Revolution research activities at the Institute (of Biophysics) were either completely interrupted or greatly curtailed." (Tien 1975, p. 627). The interruption, as Tien elaborated, was due to the participation of members of the institute in the "May 7" cadre schools, factories, or peoples' communes, thus diverting attention away from laboratory science. In the mean time, the biophysics community suffered another blow with the retrenchment of educational programs at major universities. Tien related Bei as saying that he could only discuss with Tien the biophysics program he had launched before 1966 as universities were completely shut down between 1966 and 1969. What Bei reportedly disclosed to Tien reflects the circumstances of the biophysics program at USTC.

Available biographical information on Tien suggests that he had left China before 1949, become a US citizen, and had never returned to China prior to 1973.[5] Therefore, Tien's perception of the Cultural Revolution was mostly based on what he saw and what the Chinese scientists told him during his two-and-a-half month stay in China. Given his lack of pre-knowledge with the science and society in PRC, one is left pondering: Is Tien's portrayal reliable? Does Tien's report really capture the influence of the Cultural Revolution on the Chinese biophysics community?

Recent scholarship in the history of science problematize the deliberate act to construct models of national difference while concealing pre-Nixon years of transnational scientific exchange and the more distasteful aspects of Cultural Revolution science in the 1970s US-Chinese scientific exchanges (Schmalzer 2013). It was found that the American Insect Control Delegation visiting the Chinese entomologists and entomological facilities in 1975 were engaged in what

---

[4]A separate column was devoted to discuss the impacts of the Cultural Revolution on the research at CAS and IBP respectively. See Tien (1975), pp. 626–629.

[5]Tien was known in the American scientific community for his research on experimental bilayer lipid membranes. He published 6 research articles on *Nature* between 1966 and 1970 on this subject. Tien held a bachelor degree in chemical engineering from the University of Nebraska in 1953, M.A. and Ph.D. degrees in chemistry from Temple University in 1961 and 1963 respectively. After obtaining tenure from Northeastern University between 1963 and 1966, he undertook professorship at the Membrane Biophysics Laboratory at Michigan State University in 1966 and remained there until his untimely death in 2004. See Zhang (2007), Zhuang and Wu (1997), p. 579.

Sigrid Schmalzer called "the act of comparison, the tendency to forget, and the construction of difference," as the Chinese, Chinese-American and European-American entomologists were all too keen to portray Chinese efforts in insect control as fundamentally different from corporate America. As Sigrid Schmalzer argued, "in 1975, socialist China was expected to serve as an inspirational other, its difference celebrated for what it could teach the US and the world." (Schmalzer 2013, p. 318). Since Chinese science was assumed to be different and better in delivering public goods than the US corporate-driven science, the American pressure to construct otherness and the Chinese eagerness to highlight its uniqueness and self-sufficiency necessitated downplaying and even the erasure of commonality and prior history. Schmalzer's recent article issued a felicitous warning against taking the discourse of US-Chinese scientific exchange as constituting a *prima facie* reflection of the actual status of post-Cultural Revolution Chinese science.

Compared to the US-Chinese entomological exchange, what is remarkable in the biophysical communication around the same time is the somewhat more balanced depiction of socialist biophysics in Tien's field report. On the one hand, the high level of scientific competence in the PRC certainly left a good impression on Tien:

> From my observations during my two and one-half months' visit, it appears that the scientific work being done in the PRC is of very high quality. My general impression of the laboratories I saw at the Institute of Biophysics and elsewhere is that they are staffed by dedicated and hardworking scientists and technicians (Tien 1975, p. 629).

Besides commending the work ethics of scientists and laboratory settings, Tien also praised the professional knowledge of locally-educated biophysicists and in particular their readiness to help translate technical terms in bilayer lipid membranes from English to Chinese during his lecture that was given in Chinese. It led Tien to reach the conclusion that "familiarity with recent scientific developments (in my specialty at least) is not limited to the few specialists trained abroad" and that "it was evident during the question-and-answer period following my talk that the Chinese investigators were quite up to date with the latest research in membrane biophysics done abroad." (Tien 1975, p. 630)

At the same time, Tien did not hold back from asking Chinese biophysicists about the negative impact of the Cultural Revolution on their research, nor did Chinese biophysicists appear to evade Tien's questions. Tien made no attempts to mask the disruption of the Cultural Revolution to the research and teaching of biophysics. At one point, Tien unabashedly acknowledged, "from the talks I had with scientists both at the Institute and elsewhere, my impression was that the Cultural Revolution has had many profound effects." (Tien 1975, p. 629). Tien was aware of the suspension of scientific journals and other scholarly activities during the Cultural Revolution (although he recognized that two core journals—*Scientia Sinica* and *Science Bulletin*—had resumed publication by the time of his writing) and the mandatory requirement of scientists from all ranks and all disciplines to engage in intense ideological study and practice. While Tien did not report scenarios of scientists being attacked or killed, one could sense from his writing an underlying tone of frustration and trepidation about the undesirable revolutionary effects on scientific practice.

For their part, the Chinese biophysicists responded to Tien's questions with as much details as they could disclose. They did not disguise their discontentment with the policy of compulsory participation in manual labors and curtailment of college education of biophysics. Yet neither side mentioned any biophysicist being humiliated or assaulted. It appears that neither the Chinese biophysicists nor Chinese-American biophysicists like Tien were complicit in romanticizing the science of biophysics in the first US-Chinese biophysics contact.

Tien's historic visit marked the beginning of the next era of biophysics transnationalism. Tien returned to China in 1978 and 1979 for academic exchange. In the latter visit, he met with Fang Yi—vice premier of PRC at the time—along with a command in the circle of Chinese biophysicists, the University of Chicago (*People's Daily* 1979). More important than the formal state reception is Tien's public lecture on the international updates of the teaching and research of biophysics. His talk was captured in a newly launched Chinese biophysics journal—*Progress in Biochemistry and Biophysics*.

3.4 A Biophysical Journal: *Progress in Biochemistry and Biophysics*

While Bei was busy traveling around the world for diplomatic activities in science and technology, Shen Shumin (沈淑敏) undertook the majority of the domestic management of biophysics. Shen Shumin was the second-in-command in the circle of Chinese biophysicists, next to Bei Shizhang (*PIBB* 1996). She had always been a devotee of Bei even before biophysics was instituted in 1958. She was employed as an assistant researcher in the developmental physiology laboratory in the Institute of Experimental Biology (IEB) back in 1950 when Bei acted as its founding director.[6] After IEB dissolved as the plant physiology and entomology teams left the institute in 1953, only the developmental physiology team continued to follow Bei's leadership. Shen was one of the core members who helped with the transfer of personnel and equipment from IEB in Shanghai to IBP in Beijing. After the department of biophysics was created in USTC in 1958, Bei appointed Shen as the associate head to manage most of the administrative affairs. Not only was Shen in charge of setting up the overall course outlines, development goals, and staff deployment on behalf of Bei, she was the critical link that bridged teaching resources from IBP-CAS with demands at the biophysics department at USTC. Shen acted as de facto department head as Bei Shizhang, the *de jure* head, was simply too busy to supervise the day-to-day operation of the department.

---

[6]The history of IEB as an institutional stepping-stone for the Institute of Biophysics (IBP) has been delineated elsewhere, see Luk (2014), pp. 79–87.

In 1974, Shen headstarted a journal called *Progress in Biochemistry and Biophysics* (*PIBB* 生物化学与生物物理进展) with the help of fellow biophysicists like Liang Peikuan (梁培宽) from the Academy. Shen acted as editor-in-chief for the journal between 1977 and 1987. Shen's memoirist contended that founding an academic journal was a "difficult undertaking" due to the sensitive nature of academic publishing in the shadow of the Cultural Revolution; also, the handful of scientific journals and diminishing research output made peer reviews and calls for papers an outstanding hurdle in running a journal (*PIBB* 1995).

It was estimated that there were only twenty academic journals in science and technology at the height of the Cultural Revolution in 1969. The number of scientific and technical journals in post-revolutionary China rose to 400 in 1978. This number increased by seven fold towards the end of the 1980s (Jia 2006). *PIBB* was one of the 380 journals founded between 1969 and 1978 and a major achievement of Shen. Although she was the acting chair of the biophysics department at USTC, she did not found the department or the Institute of Biophysics. *PIBB*, on the other hand, was her own brainchild rather than that of Bei Shizhang.

PIBB was not to be confused with another journal with a somewhat similar name in Chinese—*Acta Biochemica et Biophysica Sinica* (生物化学与生物物理学报). *ABBS* was founded in 1958 and run by the Institute of Biochemistry in Shanghai (presently known as the Institute of Biochemistry and Cell Biology within the administration of the Shanghai Institute for Biological Sciences). *ABBS* was initially under the helmsmanship of Wang Yinglai (王应睐), director of the Institute of Biochemistry in Shanghai and a leading biochemist in China.[7] Shen was well aware of the existence of *ABBS* when she launched *PIBB*. To avoid professional overlap, she insisted that *PIBB* would put the emphasis on "comprehensive review" (综述) and "technology and methodology" (技术与方法); the former column would introduce the latest progress and research breakthroughs from the international community while the latter would provide a platform for domestic scientists and scientific workers to exchange views with each other. These two columns became the mainstays of *PIBB*.

Establishing *PIBB* in the midst of the Cultural Revolution is a testimony to Shen's political credentials and a reflection of her application of a comprehensive approach to organizing professional communication among biophysicists. Shen's intellectual commitment to promoting biophysics as a comprehensive field of science is evident in the editorial focus of the journal, for both "comprehensive review" and "technology and methodology" were dedicated to sharing generic, interdisciplinary information that were of interest to researchers across the board.

Twenty years later, "comprehensive review" and "technology and methodology" remained core components of the journal, and stood alongside new columns such as "research reports," "research quick news and summary," "news of science and technology and information service," and "exchange of experience." (Liang 1994). The editorial office of *PIBB* highlighted the column "comprehensive review and

---

[7]For an abridged history of biochemistry in modern China, see Dong (1997), pp. 725–739.

professional papers" as "specializing in publishing the latest news in this discipline, and has become the indispensable information for preparing biology classes in various higher-education institutions, a good companion of topic selection in related science and research institutes, and predominantly a readers' favorite." (Liang 1994, p. 8). Not only does Shen's editorial philosophy stand for two decades, she also created an enduring role for *PIBB* as a principal source for teaching biology in college education. This goes along with her previous efforts to advertise and advance biophysics as "a newfound discipline in biological sciences" (*People's Daily* 1963). From 1963 to 1974, Shen moved from writing a news article to launching a new journal to promote biophysics as a biology-centered field of inquiry.

Meanwhile, one critic has implied that the accomplishment of publishing a scientific journal in 1974 was perhaps not so impressive if one considers the contents:

> When journals resumed publication around 1971, not only did most of their contents strictly address applied research, but the papers also attributed their research to the influence of Marxism-Leninism-Mao Zedong Thought. The research findings and quotations most frequently presented by scientists amounted to little more than truisms recognized as valid by any research scientist (Cao 2013, p. 128).

In light of this commentary, it seems quite necessary to delve beyond the surface of editorial ambition and evaluate the substance of the journal because the character of the journal lies in the content of its issues rather than what the editors declared it to be. Did *PIBB* only address applied research? Were the research findings and quotations published in *PIBB* contain little substance? To assess the academic value of *PIBB*—particularly in its early years—let's examine the content of articles published in *PIBB* in its maiden year of operation.

There were a total of four volumes and fifty-three articles contained in 1974's *PIBB*. In January, *PIBB* commenced the journal with an inaugural article entitled "Digging Deep into the Scientific and Technological Battlefront of the 'Criticize Lin Biao, Criticize Confucius' Campaign to Thoroughly Conduct the Socialist Revolution." (深入批林批孔把科技战线的社会主义革命进行到底). This was a propaganda article in line with the "Criticize Lin, Criticize Confucius" campaign launched in 1973 after Lin Biao's airplane crash in 1971. Although the two-page editorial had very little to do with biology or science per se, the editors consciously highlighted the inseparability between dialectical materialism and inquiry in biological sciences, through such statements as: "Biological sciences have always been a serious bone of contention between dialectical materialism, reactionary idealism and metaphysics." Political phraseology was tossed around to establish the ideological purity of the newfound journal. For example, the inaugural editorial pledged allegiance to "proletarian interests" by claiming that *PIBB* was determined to "rid itself of the influence of 'pure academic' perspectives of the bourgeoisie and make it into a combative journal with proletarian politics as the guiding force in order to unify politics and scientific operation, to criticize bourgeois and exploitative

ideologies of all kinds, and to facilitate the development of the scientific enterprise."
(*PIBB* 1974, p. 1)

The second article published in 1974 reported the symposium on "Criticize Lin, Criticize Confucius" co-organized by the editorial committees of *ABBS* and *PIBB*. Participants of the joint symposium reached a consensus that "both *ABBS* and *PIBB* should be a battlefield for the 'Criticize Lin, Criticize Confucius' campaign." (*PIBB* 1974, p. 3). Thus, *PIBB* and *ABBS* were joined in their efforts to publicize their loyalty to the mainstream political campaign in early 1974.

Judging from the contents of the first two articles, it would appear that one were to be disheartened by the spillover of politics into science. However, following the first two articles, the next eleven articles published in the first volume were all about biophysical science rather than revolutionary politics. The third article, titled "Research on the Relationship between Spatial Structure and Functions of Biological Macromolecules in Solution—Minutes of a Discussion Group," was the first piece in the "comprehensive review" column. This article summarized existing research and techniques for determining the backbone spatial structure of protein molecules, from optical rotatory dispersion and circular dichroism to hydrogen isotopes exchange, ultraviolet differential absorption, fluorescence, laser Raman spectroscopy, and nuclear magnetic resonance. The article then reviewed the biological significance of measuring macromolecular structure and envisioned the applied values of macromolecular research. Unlike the previous editorials of less than two pages, this article was six pages long and contained no propaganda vocabulary. There was no attribution of the macromolecular research to the influence of Marxism-Leninism-Mao Zedong Thought.

Likewise, the rest of the essays in volume one were all about biology in one way or another, including the bioactivity of enzymes; origins and development of mitochondrion; bio-holograms and acoustic holography; the measurement of low-level Beta exposure dose in radioactive fallouts; water-soluble protein injections; the digestive enzymes in *cyclophorus pyrostoma moellendorff*, a native mollusk to southern China; the experimental techniques to determine the molecular weight of protein molecules by gel filtration; the development and application of electron microscopes for biological research; the preparatory procedures of biological samples to be viewed under electron microscopes; and the preparation of water-insoluble enzyme derivatives and its application in biochemical analysis and separation. To sum up, only two out of the thirteen articles published in the first volume of 1974 did not touch on the research, instrumentation, and application of biological science.

The other three volumes in 1974 exhibited similar patterns as the first volume: the first two articles were assigned to editorials in support of the "Criticize Lin, Criticize Confucius" campaign or letters pledging allegiance to the Great Proletarian Cultural Revolution, while the rest of the articles were dedicated to discussion of advances in biophysics. Considering the fact that political commentaries made up less than twenty percent of the total length in each volume, it is reasonable to draw the conclusion that *PIBB*—even in its first year of operation—

did concentrate on disseminating news and research updates on biophysical science while paying some lip service to revolutionary rhetoric.[8]

PIBB was one of some three hundred journals that came into existence between 1969 and 1978. From the above content analysis, it is evident that not only was it permissible to launch a scientific journal during the Cultural Revolution, but more importantly, it was possible to devote the majority of the content to scientific knowledge and technical matters without jeopardizing the political credibility of the editorial committees or the founding members. Contrary to the skepticism that the resumed journalistic practice after 1971 only served to recycle political blather, it is clear that the bulk of the content of *PIBB* in 1974 concerned scientific progress rather than political gibberish.

3.5   The Biophysical Society of China

In many ways, *PIBB* continued to serve as a medium for Chinese biophysicists to exchange ideas about the foundation of their discipline in the post-revolutionary years. Subsequently, how to professionalize biophysics became a serious topic; the seeds sown by *PIBB* in the mid-1970s gave rise to a wider ongoing discussion on the professional orientation of biophysics in the post-Mao era.

The 1979 *PIBB* issue opened with a review article that considered points of convergence between physics and biology (Chen 1979). Fueled by the latest advances in molecular biology, this author recognized the current trend of seeking physical knowledge and methods to explain biological phenomena. The author listed several areas in which concepts and theories in physics have made revolutionary inroads in biophysical research such as mechanical forces in intermolecular analysis, electronic configuration and bonding structures of macromolecules, and biological application of nonequilibrium thermodynamics. Most of the efforts reviewed in this article involved pursuing physics as an analytical strategy to cast light on the molecular structures in biological systems. Unlike previous Chinese biophysicists, this author asserted that the physics-biology connection was expressed in the ways in which physics penetrated into biology, and that many biological problems relied on skillful physicists to tackle.

Biophysics as a specialty of using physics to study biology is a notion that had not hitherto taken root in China, at least not among mainstream Chinese biophysicists. During the Mao era, both Bei Shizhang and Shen Shumin adopted an explicitly biology-oriented conception as the foundation of biophysics. They recognized the methodological significance of physics for elucidating some of the mechanisms and problems in living organisms, but what they underscored was the equally important role of biologists with their knowledge of the connection between parts and wholes, individuals and systems. In 1964, Bei highlighted the cooperation

---

[8]All of the current and past issues of *PIBB* are available on www.pibb.ac.cn.

and mutual efforts of physicists and biologists as the underpinning of biophysics rather than a one-way flow of physical knowledge into the biological realm.

By 1980, Bei had slightly modified his view on the physics-biology relationship. Instead of directly appraising the relative significance of physics and biology in constituting biophysics, Bei placed the emphasis on the research missions and investigative purpose of biophysics:

> On the one hand, biophysics investigates the interplay between quantum effects, information processing, and changes in matter and energy; on the other hand, it connects these microscopic mechanisms with the higher performance of life at the macro level for further analysis and elaboration (Bei 1992, p. 238).

This is not a new concept. In 1964, he identified the inquiry of the basic properties of life as the purpose of biophysics, in which the study of the relationship between energy, matter, and information exchange was critical. However, Bei's formulation of biophysics in the post-Mao period did not consider biophysics' unique contribution to national defense or agriculture as contained in his 1964 article. Compared to 1964, Bei in 1980 was more concerned with the professional development of biophysics rather than the instrumental role of biophysics for state-led missions. Bei's view of biophysics underwent a transformation in the late 1970s from portraying biophysics as a mission-oriented service science into a more specialized profession geared toward academic audiences. His changing attitude towards biophysics reflects the overall institutional priority of the biophysics community in the late 1970s.

PIBB in this period was dominated by reports on the professional development of biophysics at both domestic and international levels. In a 1979 issue of *PIBB*, representatives from IBP-CAS gave a summary report on the sixth International Biophysics Congress that was held in Kyoto between 3–9 September 1978 (IBP-CAS 1979). In 1980, *PIBB* continued this trend with a follow-up article on the recent development of biophysics in Japan (Li 1980). In the spring issue of 1980, *PIBB* made two announcements concerning an important milestone in the domestic development of biophysics—the founding of the Biophysical Society of China. In February, *PIBB* featured the preparation meetings and symposiums leading to the inauguration of the Biophysical Society of China (*PIBB* 1980); and in June, it published a post-event update on the inauguration-cum-annual-meeting of the Biophysical Society of China (Liu 1980). What the above contents make evident is the surge of interest surrounding the professionalization of biophysics in the late seventies and early eighties.

In his analysis of the historical emergence of biochemistry in the US, Robert Kohler asserted that "organizing professional societies was the most overt political means of consolidating biochemical interests." (Kohler 1982, p. 198). Founding a professional society was political in the sense that it helped foster a sense of disciplinary identity among members by reinforcing the boundaries with other specialties and claiming a particular academic territory. The difference between China and the US is the different contexts in which the professionalization of science took place. Whereas the medical service role in the US provided an institutional incentive for medical chemists to market themselves as biochemists, Chinese biophysicists were gradually moving away from the Maoist policy of "using science to serve the people" and becoming a more full-blown specialized discipline. The question is: what enabled this transformation of biophysics in the late 1970s?

The emergence of professional societies in science and technology in post-Mao China was facilitated by the policies of economic liberalization introduced by Deng Xiaoping in 1978. Deng's administration was marked by a profound optimism surrounding the engine of market reforms to fuel the Four Modernizations program that he advocated. In the scientific community, the beginning of Deng's era was sometimes called the "springtime for science" to underline its more libertarian attitudes to scientific affairs compared to the earlier regimes. While attempts to claim elite status for scientists or specialized knowledge were greeted with skepticism in Mao's China, such endeavors were permissible and even desirable in the eyes of Deng Xiaoping. Since science is a pillar of the Four Modernizations, advancing the professional interests of scientists is considered acceptable as long as the ultimate goal of science is to realize the dream of the Four Modernizations.

For the Chinese biophysicists, this presented a window of opportunity to further strengthen the disciplinary coherence of biophysics. In his opening speech at the inauguration of the Biophysical Society of China, Bei Shizhang declared: "the founding of the [Biophysical] Society opens a new page in the enterprise of biophysics in our country. It will certainly catalyze the professional development of biophysics." Organizing a professional society departs from the previous model of "mission drives discipline, discipline facilitates mission." In post-Mao China, Chinese biophysicists were less concerned with facilitating state-led missions than cultivating a collective sense of professional identity.

The Biophysical Society of China was organized by a group of biophysicists in the late seventies with the goal of strengthening the professional status of biophysicists. According to the Society's yearbook, the formation of the Society was intended to "unite all scientists and workers in biophysics in order to facilitate the development of biophysics in China." In its self-portrait, the Society was in nature an organization to protect and expand "academic specialty." (The Biophysical Society of China 2011). Before 1979, biophysicists were a division in the Physiological Society. The founding of the Biophysical Society in 1980 marked a professional boundary separating biophysics from physiology in order to further the professional interests of the biophysicists.

Since its inception in 1980, the Biophysical Society of China has become a platform for learned activities and professional development. In addition to its founding mission and philosophy, the Society pushed for heightened international collaboration and exchanges. When the Society was founded in 1980, it was not a member of the International Union for Pure and Applied Biophysics (IUPAB). After several years of preparation and negotiation, the Biophysical Society of China became a member of IUPAB in 1984. Since then, the Society has been a regular participant in its annual meetings around the world. In 2011, the Society was granted the right to convene the 17th IUPAB annual meeting in Beijing. The right to host the IUPAB meeting was likened to the right to host the Olympic Games in the biophysical circle (The Biophysical Society of China 2011). By hosting the IUPAB meeting in China, it was hoped that big events like this could catalyze and accelerate the professional development and infrastructure improvements necessary to ensure that Chinese biophysicists were among the international community of biophysicists.

This strategy to align Chinese scientists with the international realm of scientific players is a reflection of Deng's "open-door" policy. It was a move away from the isolationist policy in the previous regime by embracing opportunities opened up by the neo-liberal economic principle. Also characteristic of Deng's "open-door" policy is the increased leeway given to the dissemination of foreign information. Greater access to international updates in science and technology sharpened Chinese scientists' sensitivity to the organizational infrastructure of science and technology abroad. The growing appetite for knowledge about the development of biophysics in foreign countries was reflected in the content of *PIBB* in the early 1980s. Readers of *PIBB* were exposed to an increasing number of reports on the research, teaching, and organizational practices of biophysics in other countries.[9]

3.6 Concluding Remarks

This chapter illustrates the institutional structure of biophysics in areas of research, teaching, journalistic, and professional activities. Prior to the founding of *PIBB* and the Biophysical Society of China, biophysics was taught at the USTC and practiced by researchers at the IBP-CAS from 1958 to the eve of the Cultural Revolution.

During the Cultural Revolution, Chinese biophysicists discontinued the research and teaching of biophysics but as a group they were quite successful in shielding themselves away from revolutionary abuse. The significant question is: Why were biophysicists apparently successful in insulating themselves? One reason was that leading biophysicists, like Bei Shizhang, were in good political standing. This is reflected in Bei's ambassadorial role in the first Chinese delegation traveling to the US and Western Europe beginning in 1972. Secondly, the objectives of the biophysics leadership were not significantly different from those of the political arena at large. Biophysicists aiming at discipline-building are not at odds with the interests of the state opting for nation-building.

The changing leadership from Mao to Deng presented a changing political context for Chinese biophysicists and their discipline-building project. Since biophysics, just like most other scientific disciplines, was required to serve the national needs during the Maoist regime, much of the disciplinary building of biophysics was conducted under the aegis of the centralized state to advance "nation-building." After Mao, revolutionary heroism was downplayed among the policy-making

---

[9]See the 1980 issue of *PIBB* on www.pibb.ac.cn.

authority in Deng's cabinet. In this context, an increased professionalism was symptomatic of the institutionalization of biophysics during the post-Mao years. The new journal and new professional society promoted a new sense of occupational identity among the biophysicists by disseminating new forms of specialized knowledge (that was largely detached from revolutionary rhetoric) and international exchange of increasing breath and depth in the recent years.

References

Apter D, Saich T (1994) Revolutionary discourse in Mao's republic. Harvard University Press, Cambridge

Bei S (1992) Selected writings of Bei Shizhang (SWB). Zhejiang Science and Technology Press, Hangzhou

Brock D (2013) The people's landscape: mr. science and mass line. In: Brock D, Wei CN (eds) Mr. science and chairman Mao's cultural revolution: science and technology in modern China. Lexington Press, Lanham, pp 41–118

Cao C (2013) Science imperiled: intellectuals and the cultural revolution. In: Brock D, Wei CN (eds) Mr. science and chairman Mao's cultural revolution: science and technology in modern China. Lexington Press, Lanham, pp 119–142

Chan P (1992) China. In: Wielemans W, Chan PC-P (eds) Education and culture in industrializing Asia. Leuven University Press, Belgium, pp 39–110

Chen RS (陈润生) (1979) Several biophysical problems (几个生物物理学问题). PIBB 6:35–40

Communication with the People's Republic of China (CSCPRC) (ed) (1986) A relationship restored: trends in U.S.-China educational exchanges, 1978–1984. National Academy Press, Washington, DC

Dong GB (董光璧) (1997) History of science and technology in modern and contemporary China (中国近现代科学技术史). Hunan Education Press, Changsha

Esherick J, Pickowicz P, Walder A (2006) The Chinese cultural revolution as history: an introduction. In: Esherick J, Pickowicz P, Walder A (eds) The Chinese cultural revolution as history. Stanford University Press, California, pp 1–28

Fan HY (樊洪业) (ed) (1999) A historiography of the Chinese Academy of Sciences 1949–1999 (中国科学院编年史, 1949–1999). Shanghai Education Press, Shanghai

Gao Y (1987) Born red: a chronicle of the cultural revolution. Stanford University Press, California

Institute of Biophysics-Chinese Academy of Sciences (IBP-CAS) (1979) A summary report on the sixth international biophysics congress. (第六届国际生物物理会议概况介绍). PIBB 6:17

IBP-CAS (2008) Flying dogs in the sky: a documentation of Chinese biological experimental rockets (小狗飞天记：中国生物火箭试验纪实). Science Press, Beijing

Jia X (2006) The past, present, and future of scientific and technical journals of China. Learned Publishing 19(2):133–141

Kohler R (1982) From medical chemistry to biochemistry: the making of a biomedical discipline. Cambridge University Press, Cambridge

Lampton DM (1977) The politics of medicine in China: the policy process, 1949–1977. Westview Press, Boulder

Lewis J, Xue LT (1994) China's strategic seapower: the politics of force modernization in the nuclear age. Stanford University Press, California

Li H (李晔) (1980) Development of biophysical research in Japan (日本生物物理学研究的发展) PIBB 7:78–80

Li CZ (李成智) (ed) (2005) A draft history of the development of space technology in China, vol. 1–3 (中国航天技术发展史稿, 上中下册). Shandong Education Press, Jinan

Liang PK (梁培宽) et al (1994) Fostering cooperation, marching forward—In commemoration of the twentieth anniversary of Progress in Biochemistry and Biophysics (加强合作, 继续前进—纪念《生物化学与生物物理进展》创刊二十周年). PIBB 21:7–8

Liu R (刘蓉) (1980) Inauguration of the biophysical society of China in Beijing coincided with the third annual meeting of biophysics in China (中国生物物理学会成立大会在京召开, 第三届全国生物物理学学术会议同时举行). PIBB 6:88

Luk CYL (2014) Biophysics, rockets, and the state: the making of a scientific discipline in contemporary China. Ph.D. Diss, Arizona State University

Luk CYL (2015) Building biophysics in mid-century China: the University of Science and Technology of China. J Hist Bio 48(2):201–235

MacFarquhar R, Schoenhals M (2006) Mao's last revolution. Harvard University Press, Cambridge

Miller HL (1996) Science and dissent in post-mao China: the politics of knowledge. University of Washington Press, Seattle

Nie RZ (1988) Inside the red star: the memoirs of Marshal Nie Rongzhen. New World Press, Beijing

Peng JQ (彭洁清) (2002) His passion for spaceflight: boundless love and yearning (航天情: 永远的眷恋). Tsinghua University Press, Beijing

People's Daily (1963) Biophysics–a newfound discipline in biological sciences (生物科学中的一门新兴学科—生物物理学). People's Daily, 6 August

People's Daily (1972) Embarking on a friendly visit to the United Kingdom, Sweden, and Canada: Bei Shizhang leads a Chinese science delegation departing from Beijing, Guo Moruo, Liu Xiyao, Zhang Wenjin, Wu Youxun, Zhou Peiyuan etc. come to the airport to see them off (前往英国、瑞典、加拿大进行友好访问:贝时璋率中国科学家代表团离京,郭沫若、刘西尧、章文晋、吴有训、周培源等到机场送行). People's Daily, 7 October

People's Daily (1979) Fang Yi received (Chinese) American scholars Nien-Chu Yang and Hsin-Ti Tien (方毅会见美籍学者杨念祖、田心棣). People's Daily, 30 August

PIBB (1974) Digging deep into the scientific and technological battlefront of the 'Criticize Lin Biao, Criticize Confucius' campaign to thoroughly conduct the socialist revolution (深入批林批孔把科技战线的社会主义革命进行到底). PIBB 1:1–4

PIBB (1980) The biophysical society of China will inaugurates in the second quarter together with the third academic symposium (中国生物物理学会成立大会将于二季度召开, 第三次学术讨论会同时举行). PIBB 6:27

PIBB (1995) An indefatigable journal founder delights in the mission of education—celebrating the eightieth birthday of professor Shen Shumin (办刊育人, 乐此不倦—沈淑敏先生八十寿辰致庆). PIBB 22(1):2–4

PIBB (1996) In woeful mourning of professor Shen Shumin (沉痛悼念沈淑敏先生). PIBB 23 (6):482

Pyne SJ (2011) Voice and vision: a guide to writing history and other serious nonfiction. Harvard University Press, Cambridge

Rapoport Y (1991) The doctor's plot of 1953: a survivor's memoir of Stalin's last act of terror, against jews and science. Harvard University Press, Cambridge

Schmalzer S (2013) Insect control in socialist China and the corporate United States: the act of comparison, the tendency to forget, and the construction of difference in 1970s US-Chinese scientific exchange. Isis 104(2):303–329

Science Times (科学时报) (ed) (1999) May history remember them: chinese scientists and 'Two Bombs, One Star' (请历史记住他们: 中国科学家与'两弹一星'). Jinan University Press, Guangzhou

Solomone S (2013) Space for the people: china's aerospace industry and the cultural revolution. In: Brock D, Wei CN (eds) Mr. science and chairman Mao's cultural revolution: science and technology in modern China. Lexington Press, Lanham pp 233–250

Song J (宋健) (ed) (2001) Biographies of pioneers of 'Two Bombs, One Star', vol 1 and 2 ('两弹一星'元勋传-上下卷). Tsinghua University Press, Beijing

Song MY (宋敏毅) (2005) An interminable passion for space: the consort of 'Two Bombs, One Star' medalist Yao Tongbin visits Tian Yi middle school (悠悠航天情:'两弹一星'功臣姚桐斌的夫人彭洁清在天一), Xi Shan Jiao Yu, 7 December

The Biophysical Society of China (2011) Collection of essays in commemoration of the thirtieth anniversary of the biophysical society of China (中国生物物理学会成立三十周年纪念文集). The Biophysical Society of China Press, Beijing

Tien HT (1975) Biophysical research in the People's Republic of China. Biophys J 15(6):621–631

Wang GY (2010) Bei Shizhang: a biography. Science Press, Beijing

Xinhua News (2005) Former residence of Yao Tongbin, a 'Two Bombs, One Star' pioneer, is open to the public after renovation ('两弹一星'元勋之一姚桐斌故居修复对外开放) Xinhua News, 13 November

Xu LY et al (eds) (1982) Science and socialist construction in China. M E. Sharpe, New York

Yang GY (杨国宇) (2000) Information of local literature: diary no. 24—a General's diary during military control I: 1967–1969 (民间语文资料:日记024号—将军军官日记, 上:1967–1969) Frontiers (天涯) 5:1–8

Ying YM (应幼梅) (1992) The life, work and thought of Professor Bei Shizhang (贝时璋教授的生活、工作和思想). In: SWB. Zhejiang Science and Technology Press, Hangzhou, pp 1–39

Zhang J (张钧) (ed) (1986) Aerospace industry in contemporary China (当代中国的航天事业). Chinese Social Science Press, Beijing

Zhang JW (张建伟), Deng CC (邓琮琮) (1996) Chinese academicians (中国院士). Zhejiang wen yi chu ban she, Hangzhou

Zhang L (张藜) (2009) The economic livelihood and social status of scientists: 1949–1966—The case of the Chinese academy of sciences (科学家的经济生活与社会声望:1949–1966年——以中国科学院为例). Studies of the contemporary history of China 中国当代史研究, Jiuzhou Press, Beijing, pp 104–142

Zhang ZR (章宗穰) (2007) Introducing advances in planar lipid bilayers and liposomes series—in commemoration of the 45 anniversary of the publication of research in artificial bilayer lipid membranes (《平面类脂双层和脂质体研究进展》丛书简介—纪念人工类脂双层膜研究工作发表45周年). Chem Sens 化学传感器 27(1):66–68

Zhu QS (朱清时) (ed) (2008) A draft edited history of the University of Science and Technology of China. University of Science and Technology of China Press, Hefei

Zhuang YL (庄炎林), Wu J 伍杰 (eds) (1997) A dictionary of overseas Chinese and foreign affairs (华侨华人侨务大辞典). Shandong Friendship Press, Jinan

Chapter 4
Building Biophysics Through Launching Sounding Rockets

Abstract This chapter explores how leading Chinese biophysicists chose to advance their discipline through launching biological sounding rockets. The participation of biophysicists in Chinese space and rocketry program, *liangdan yixing*, signified the ways in which leading biophysicists found ways to incorporate the appropriate technologies into specific institutional contexts. What this chapter discusses is how the technology that was available shaped what the individuals with knowledge of biophysics could do to meet the needs of their institutional benefactors. Ultimately meeting this need for their expertise is what granted the biophysicists the resources to build and develop the discipline. The specifics of the construction of biophysics in China are examined in this chapter through the history of the launching of the biological rockets between 1958 and 1966. The guiding questions of this chapter concern how biophysicists negotiated their roles in the space program as well as what the space mission meant to biophysics as a discipline.

Keywords Biological sounding rockets · Chinese space and rocketry program · Chinese T-7 research rockets · Institute of biophysics · Missile and bomb research

The last chapter discussed the institutional component of biophysics in contemporary China. In the late fifties, not only was there an evolving intellectual interest clustering around biophysics, there was also an increasing movement towards the institutionalization of biophysics. In 1958, a research institute and a department of biophysics were founded; by 1980, a scientific journal and a professional society for biophysics were launched. Yet these milestone events are only meaningful if the people within the institution made use of these structural resources; the case being that without substantial research content, the institutional infrastructure is nothing more than an embellished shell. In light of this, it is instructive to ask: What specific scientific and research activities were undertaken? How did biophysicists make use of this institutional platform to further its disciplinary cause?

Institutional patronage for biophysics is not uncommon in other places at some points. In the same year as biophysics was instituted in China, the American biophysicist Francis Schmitt organized a month-long national conference on "the

© The Author(s) 2015
C.Y.L. Luk, *A History of Biophysics in Contemporary China*,
SpringerBriefs in History of Science and Technology,
DOI 10.1007/978-3-319-18093-9_4

study program in biophysical science" at Colorado with generous funding from the NIH. His ambition was to solicit a national consensus on the disciplinary programming of biophysics. It did not turn out as well as he had wished. Rasmussen (1997a) has diagnosed some of the underlying causes. One of the problems is that the institutional commitment to biophysics did not last long in the US. The programmatic collapse of biophysics in the US testified to the importance of institutional support to buttress a fragile discipline with flexible research areas and lucrative technological promise. If biophysics was to be put on a permanent footing as a fully mature discipline, its leaders had to find ways to incorporate the appropriate technical advances into the specific institutional contexts (Rasmussen 1997b).

What this chapter discusses is how the technology that was available shaped what the individuals with knowledge of biophysics could do to meet the needs of their institutional benefactors. Ultimately meeting this need for their expertise is what granted the biophysicists the resources to build and develop the discipline. The specifics of the construction of biophysics in China are examined in this chapter through the history of the launching of the biological rockets between 1958 and 1966. In documenting how Chinese biophysicists built the discipline of biophysics through launching the rockets, two sets of questions were brought to the forefront: The first is how biophysicists interacted with other experts in the Fifth Research Academy of the Ministry of National Defense (aka the Fifth Academy)— the chief unit responsible for putting together rockets and missiles in socialist China. The early history of the Fifth Academy reveals how government bureaucracy restructured the academy in order to create a workspace for rocket and missile scientists. This introduces the question of how biophysicists positioned themselves in the Fifth Academy when it was populated by rocket engineers, and also of how they combined discipline building with the politics of scientific collaboration.

The second set of questions is how the Chinese space program contributed to the institutionalization of biophysics. Space and rocketry programs are usually mission-centered whereas scientific disciplines are more research-and-teaching oriented. A mission is a series of short-term operations while research and teaching involve long-term planning. So in what ways did the space program, directly or indirectly, enable the disciplinary growth of biophysics in socialist China?

In other words, the two overarching questions of this chapter concern how biophysicists negotiated their roles in the space program as well as what the space mission meant to biophysics as a discipline.

4.1 Mission 581 and "Two Bombs, One Star"

The year 1958 was an *annus mirabilis* for Chinese biophysicists. The formation of a biophysics research institute and a biophysics department in this year coincided with one of the most important trans-disciplinary, cross-sectional missions that ever took place in the history of socialist China.

Instigated by the Soviet launch of Sputnik in 1957, top leaders at the Academy of Sciences such as Zhu Kezhen (竺可桢), Qian Xueshen (钱学森), and Zhao Jiuzhang (赵九章) proposed that the Academy should design a satellite project that went along with the military projects on atomic and hydrogen bombs. This idea was conveyed to the Party central committee by Zhang Jingfu (张劲夫), then vice president of the Academy. The heat from Sputnik triggered a surge of policy interests in artificial satellites. On 17 May 1958, Chairman Mao announced at the second session of the Eighth Annual Meeting of the National Congress: "we are going to develop artificial satellites." (Wang 2010a, Wang 2010b). Within the same month, the state secretariat sanctioned the artificial earth satellite plan. This was the first artificial satellite plan put forth by the Chinese government, and CAS was the major supplier of expertise to execute this project. The project was known as "Mission 581" to denote the fact that it was the mission of first priority devised in 1958.

An artificial satellite was prioritized as the key mission of the Academy in 1958. An integral part of the satellite project was to develop sounding rockets (探空火箭). Sounding rockets are the most basic type of missile-based research vehicles. They can carry instruments and animals to the atmospheric and ionospheric strata but cannot penetrate into outer space. Sounding rockets are flown on missions of short duration and are not intended to reach orbit. Compared to the more powerful and advanced launch vehicles such as space shuttles, sounding rockets are not nearly as high tech, but they do have their advantages. They are much cheaper and less risky for geophysical and biophysical experimentations at high-altitude. The relatively low cost and low risks make them a popular instrument for upper atmospheric research. Accordingly, before 1968 the National Aeronautic and Space Administration (NASA) recommended that new instruments designed for satellite flight first test themselves on sounding rockets (Corliss 1971). In other words, sounding rockets are the prerequisites for satellite flights.

In China, Deng Xiaoping was in favor of experimenting with sounding rockets in space research rather than lavishing resources on the more expensive and complex technology of earth satellites. This was captured by a slogan popularized by Deng, "turning thighs to calves, satellites to sounding (rockets)" (大腿变小腿, 卫星变探空).[1] In view of the material shortages in the late 1950s, the Academy focused on launching two types of sounding rockets, namely the meteorological sounding rockets (气象探空火箭) and the biological sounding rockets (生物探空火箭). The Institute of Biophysics was an important participant in the biological sounding rockets program, along with the Academy of Military Medical Sciences and the Chinese Academy of Medical Sciences. The entire project was supervised by the Fifth Academy under the leadership of the famous rocket engineer Qian Xuesen (Harvey 1998).

This is our first glimpse into the role of Chinese biophysicists in the early history of the Chinese space program. Biophysicists were recruited into the bio-rocketry

[1]"Turning thighs to calves" is a metaphor that means something akin to "shortening" and in this context refers to reducing expenses in the rocket decision-making process. See Song (2001).

program in Mission 581, which was never just a civilian science project undertaken by the Academy; rather, it was meant to be a joint civilian-military operation from the very beginning. This was evident in the close relationship between the Academy and various military departments, most notably the Fifth Academy. Together with the hydrogen and atomic bombs, Mission 581 was part of the large-scale techno-military project called "Two Bombs, One Star" (*Science Times* 1999).

Rather than seeing the military participation in the project as a form of intervention, the constituency of Mission 581 reflects the indivisibility of scientific merit and military value. Technologically speaking, missiles and bombs are inseparable from satellites because a reliable carrying vehicle is the foundation for lifting bombs and satellites off the ground. The missile is the major apparatus for propelling anything across a long range within a short time. Although atomic bombs, hydrogen bombs, missiles, and satellites are inter-related, developing missiles is the first step. But since the missile is primarily a weapon, it belongs to a strategic area under military supervision. The central unit overseeing the Chinese missile program was the National Defense Science and Technology Commission (NDSTC), which was headed by Marshal Nie Rongzhen (聶榮臻) under the jurisdiction of the Central Military Commission. John Lewis and Xue Litai have related the role of the military-industrial complex in China's nuclear weapon programs. Lewis and Xue (1988) told the story of Chinese bomb-making from the perspective of international politics of arms control as well as about the meanings of missiles and nuclear weapons to China's defense modernization. What Lewis and Xue underlined was the strong military implication of missiles and nuclear weapons.

Most Chinese military officers and government ministers welcomed the strategic acquisition of a nuclear arsenal for the sake of national interests and self defense. The American nuclear threat was a fundamental driving force behind this strategic decision. Lawrence Freedman once remarked, "no country had been closer to nuclear attack than the Chinese since Hiroshima and Nagasaki were destroyed" (Freedman 1981, p. 276). It is therefore understandable that many high-rank revolutionary leaders embraced nuclear weapons out of concern for national security.

In the meantime, the focus of this chapter is not the military commanders and government bureaucrats as featured in the narrative of Lewis and Xue, but the scientists in the missile–cum–satellite program. What did the scientists at the Academy think of the strong military presence in Mission 581? The geophysicist Zhao Jiuzhang, who was in charge of launching the geophysical sounding rockets, emphasized the military value of satellites. In a corresponding letter addressed to Zhou Enlai in 1964, he wrote "almost all satellites are related to national defense. The CAS planning proposals for developing China's artificial satellites should also stress the military application of satellites." (Li 2005a, p. 611)

As for other scientists' views on the civilian-military nexus, it was reported that almost two thirds of all the personnel at the Academy were mobilized to work for "Two Bombs, One Star" at one point (*Science Times* 1999). How did these scientists negotiate their roles and interests in this military-coordinated program? Was there any jurisdictional dispute between specialists and commanders from different agencies?

Civilian-military conflict in space programs is quite common. Space science and technology is a sophisticated feat of engineering that requires multidisciplinary cooperation between experts, technicians, and administrators. Friction is mostly unavoidable in close interactions among people with heterogeneous backgrounds and values. Scientists are not accustomed to military styles of operation while military personnel are not trained to consider long-term research questions beyond their assigned tasks and missions. Both technical and operational factors are necessary to successfully execute a space program, for a sustainable space program has to carry out both high-risk missions in a timely and safe manner while also transferring some of the military benefits for civilian use. Striking a balance between these multiple factors can be difficult. In what follows, I will attempt to elucidate the civilian-military context in which Chinese space program was conducted.

4.2 Contextualizing "Two Bombs, One Star"

The private negotiation between Song Renqiong (宋任穷) and Zhang Jingfu exposed some of the dilemmas in civilian-military cooperation in "Two Bombs, One Star." Song was the bureau chief in charge of China's nuclear industry, first called the Third Ministry of Machine Building (Third MMB) between 1956 and 1958, then the Second Ministry of Machine Building (Second MMB) after 1958. Zhang was the vice-president of the Academy at the time and was a pivotal figure in Lewis and Xue's account (Lewis and Xue 1988). Before the Academy came into existence, Zhang had worked under Song in Communist guerilla battles. During the "Hundred Regiments Campaign" in 1949, Zhang was the deputy commissar of the Anhui province while Song was the secretary of the provincial committee and the political commissar. Not only were they colleagues, they were comrades-in-arms and close friends. When "Two Bombs, One Star" was underway in 1958, Song called Zhang's office one day saying that he wanted to make a home visit to see him. It made Zhang uneasy to have his "big brother" visit him; Zhang suggested that he should go see Song instead. But Song insisted he had to come to Zhang, not the other way round. Zhang then understood that Song approached him for soliciting support from the Academy for the Second MMB. When Song saw Zhang, he dashed over and clutched Zhang's hands, "Jingfu, this is too important a matter. You have to help! I had hoped other departments would contribute, but the major support comes from the Academy!" (*Science Times* 1999, p. 24). They reached an agreement to transfer the Institute of Nuclear Science originally established in 1950 under the Academy over to the Second MMB but maintained its official title as "INS-CAS" in the public domain. INS-CAS was under the dual leadership of the Academy and the Second MMB. This arrangement was made to better align the nuclear expertise with the military needs of making weapons, as it represented a military overshadowing of civilian control of nuclear science. Zhang was well aware of the political overtones as he announced to scientists at INS, "After INS was handed over to the Second

MMB, it is not necessary for you to come to CAS meetings. Go to the meetings at Second MMB instead." (*Science Times* 1999, p. 25)

But Zhang did not always give way to military encroachment. When Liu Youguang (刘有光), commissar of the Fifth Academy, invited Zhang to sit on the party committee of the Fifth Academy, Zhang resisted, realizing that Liu's purpose was to facilitate a further brain drain from the Academy to the defense sector. Zhang opined, "I am with the Academy. How can I go to the army? This is not going to work." (*Science Times* 1999, p. 35). Since Marshal Nie also rejected Liu's motion, Liu had no alternative but to drop the idea. Accordingly, Zhang sought to balance civilian-military equilibrium by invoking the analogy of "walking-on-two-legs:"

> I propose (the strategy of) walking on two legs. On the one hand we cooperate with the Fifth Academy; at the same time we carry out our own project here at CAS …Nie agrees with this suggestion. CAS should also carry out missile research because most experts—except the elites that have been transferred to the Fifth Academy—remain in CAS. There are many research institutes in CAS. We have a comprehensive team. Therefore China decides to pursue missiles with two legs (i.e. from two avenues): the first one is the Fifth Academy. It is the central bureau specialized for this task with generous support from the state; another is CAS. We also do research and explorative work. We have our experimental sites.

The "walking-on-two-legs" trope in Zhang's quote is not a random metaphor. It is a widespread figure of speech in socialist China that is employed for striking a balanced relationship between binary opposites such as theory and practice, research and production, experts and amateurs. It was inscribed in a field report put together by a team of radical American scientists and teachers visiting China in the early 1970s who witnessed the grassroots infiltration and mobilization of national campaigns for "the benefit of the people." (Science for the People 1973)

Although some socialist enthusiasts were disillusioned with the deceptively optimistic accounts, not all aspects of socialist science were misleading. The dual civilian-military coordination and conflict in "Two Bombs, One Star" exemplifies the dialectical mode of incorporating practitioners from separate workplaces into a cooperative site while retaining their independence. The "walking-on-two-legs" metaphor encouraged scientific integration in a dialectical fashion. It was "dialectical" in the sense that it manifested the principle of the unity of opposites in dialectical materialism (Graham 1974).

The dialectical relationship between scientific organizations such as the Academy and military agencies such as the Fifth Academy or the Second MMB is the backdrop against which the biophysicists' contribution to the space program can be critically assessed. It is imperative to remember that like all space endeavors, the Chinese space program is composed of a miscellaneous group of practitioners and politicians. It is an arena of political struggle where at times one side would cave in for better overall coordination but at other times it strove for harmony and order by synthesizing opposing forces. Biophysicists represented the civilian force from the Academy. Examining their interactions with colleagues in the military service such as flight surgeons and rocket designers can reveal more intricate details about the role and responsibility of biophysicists in the Chinese space program.

4.3 Comparative Politics of Civilian-Military Alliance in Space Missions

Space operations are made up of an eclectic group of experts in which nuclear physicists and rocket engineers occupy a more central and exalted place than biologists and life scientists. This is hardly surprising given the fact that a large chunk of a space mission is attached to technical performance. The majority of this responsibility falls upon the shoulders of engineers and physical scientists. Life scientists are usually incorporated into the space program for the purpose of supervising manned operations in space. But this "life" variable has to wait until the "mechanical" variable is tackled. In other words, only after the aircraft can be successfully lifted off the ground can anyone consider how to put living things into the capsule.

Since engineers and technicians tend to dominate space missions, the role of biomedical scientists at the early stage is usually marginalized. The adversarial relationships that exist between biomedical specialists, physical scientists and administrative officers in the NASA Manned Space Program has been noted by Pitts (1985). In NASA, life science as a discipline was perceived as subordinate and secondary to the agency's major space programs from the very beginning. NASA's first administrator T. Keith Glennan created the biomedical team as an adjunct to the Space Task Group rather than as an independent unit. After that, there was no direct effort to convert the adjunct division of life sciences into a permanent office, nor was there any political will to appoint a life scientist to a high-level position of deliberation.

The marginalization of biomedical science led to internal factionalism among clinicians and technicians in the Space task force. As Pitts suggested, "the subordination and decentralization of the life sciences…would preclude the interaction among biologists, medical scientists, and the clinicians that is normal in biomedical research setting…many scientists questioned NASA's ability to provide adequate biomedical support for manned spaceflight." (Pitts 1985, p. x). As a result, NASA decision-makers reluctantly set up a Life Sciences Advisory Committee in preparation for the Mercury project; yet a staff report concluded after the Mercury operation that "NASA was underestimating the importance of biomedicine." (Pitts 1985, p. 16)

The domination of engineers in NASA top management is a primary reason for the lack of significance assigned to life science in the space program. But it also has a lot to do with the turf conflict over space biology (or space life sciences) between NASA and the US Air Force. In many ways, the Air Force was ahead of NASA in its acquired capabilities of space biomedicine, in part because the Air Force antedated NASA in developing aviation medicine after WWII. As Maura Mackowski argued, "space medicine came into being as the product of several factors: first and foremost, the development of aviation medicine into a recognized professional specialty." (Mackowski 2002, p. 2). The boundary between aviation medicine and aerospace medicine is not so much a scientific boundary, but rather a political one.

During the interwar period, aviation medicine was under the exclusive control of the military services as aviation medicine was inseparable from combat medicine and flight surgery.[2] In contrast, space biomedicine lies somewhere along the military-civilian continuum. Essentially, the Air Force accounted for a sizable part in the professionalization of aerospace medicine in the US. After NASA was founded in 1958, the Air Force continued to gain preeminence on biomedical expertise in orbital and sub-orbital spaceflights. The hesitation of NASA administrations to authorize an official and coherent life science program within the agency aligned with the Air Force's political preference. Accordingly, the Air Force took concrete steps to prevent NASA from strengthening its own biomedical facilities. For example, the Air Force teamed up with their lobbyists in the Congress to deny NASA funding for building an in-house biomedical program. As Pitts related, "a major expansion of in-house capabilities in the life sciences (at NASA) ran directly counter to the aspirations of the Air Force." (Pitts 1985, p. xi). NASA did manage to create a Division of Life Sciences in the mid-1970s but it "did not lead to a truly integrated life science program." (Pitts 1985, p. 182). "How biology fits into and interacts with larger systems" remained a challenge in NASA well into the 1980s if not later (Pitts 1985, p. 208).

The clash between NASA and the Air Force crystallized the abrasive relationship between civilian agencies and military services in pursuing space-related biosciences in the US, with the problem being further intensified by the historical contingencies that shaped the creation of NASA. In spite of Eisenhower's rhetoric of separating civilian space exploration from military activities, NASA was founded primarily not for scientific exploration but with the explicit goal to "beat the Soviets." (Wang 2008). It was international politics rather than domestic scientific interests that gave rise to NASA. For this reason, John Pitts was conscious of the fact that "NASA was not primarily a science agency; it demanded a form of organization and management that reflected space program objectives and capabilities rather than scientific priorities alone." (Pitts 1985, p. 176)

That NASA was a non-scientific construct disguised as a civilian agency complicated the recruitment of life scientists in its civilian space programs. If NASA was an ordinary scientific organization like the NSF or the NIH, efforts to launch a life-science-oriented program would have met with less resistance. However, NASA was not an agency designed to advance science for the sake of science. Hence, how to coordinate the biomedical team in a decentralized political culture while maintaining the agency's priorities to fight against the Soviets remained a major problem.

In the previous section, I revealed some of the dialectical relationship between the civilian unit and defense sectors in China. It is noteworthy that the history of

---

[2]For example, the first professor of aviation medicine, Hubertus Strughold maintained that the distinction between aeronautical and astronautical flights was as artificial and misleading as that between space and atmosphere. For Strughold, space medicine is a logical extension of aviation medicine. See "Hubertus Strughold and Aviation Medicine in Germany, 1927–1945," in Mackowski (2002), pp. 79–140.

CAS is longer than that of NASA. Unlike NASA, CAS was created not with the doctrine of international combat in mind but with the objective of promoting domestic science and technology for national modernization. Thus, there had already been a cadre of scientists in basic research when the grandeur of Sputnik stunned the world in 1957. It is true that the overall scientific capabilities and facilities in China were not on a par with those in the US or the USSR in the 1950s but Chinese scientific brainpower was by and large concentrated in one centralized agency. Consequently, the Second MMB from the Ministry of Defense had to recruit and borrow specialists from CAS—just as Marshal Song insisted on approaching Zhang, the deputy director of CAS, and not the other way around— because CAS was the national storehouse of scientific experts and expertise.

While the transfer of space-related scientists flowed from the civilian agency to the military sector in China, the reverse happened in the US. The Air Force was the major employer of specialists in the sciences pertaining to airpower and aviation services in post-war America. The monopoly of aviation medicine by the Air Force (and the Army to a lesser extent) stemmed from the triumph of the US as an air super-power in the Pacific battlefield. It showcased the military prowess of the US in the international aerosphere when aviation medicine in Japan (and Asia in general) was well behind that of Germany or the US during WWII. The predominance and pre-existence of the Air Force outstripped NASA in biomedical capabilities for space operations. NASA was founded "in Sputnik's shadow"[3] while the military superintendence overshadowed the civilian control of space research. This specific civilian-military organizational arrangement was as much a historical happenstance as an institutional outcome of the decentralized responsibilities for undertaking R&D activities in the decentralized public system of accountability in the US political culture.

Existing evidence suggests that a decentralized political system is not always conducive to optimize federal coordination. When there is a dire need for directing instrumental actions, a central unit of authority has the advantage of mobilizing resources and enforcing discipline in large-scale projects that require seamless cooperation among multiple agencies. The subordination of life scientists in NASA's space operation illustrates this point. In the late fifties and early sixties, US biomedical capabilities for space operations were dispersed among different agencies and the related activities in aviation physiology were largely uncoordinated and/ or under-coordinated. Dr. Randolph Lovelace II, a reputable flight surgeon who chaired the pilot selection committee for Project Mercury, urged for "a coordinated national program of research in space biology and medicine." (Pitts 1985, p. 10). Building on Lovelace's opinion, an external advisory committee under the chairmanship of Dr. Seymour Kety from the NIH issued a report pushing for a pro-integration view in managing NASA's biomedical programs. The "Kety's report" in 1960 recommended a "maximum integration of the personnel and facilities

---

[3]"In Sputnik's shadow" is the book title of Wang Zuoyue. Wang used the metaphor to convey his thesis of a new storm of technological enthusiasm in Sputnik's aftermath. See Wang (2008).

applicable to the space-related life sciences in the military services and other Government agencies." (Link 1965). The advisory report highlighted the lack of coordination between civilian agencies and military services in pursuing space biosciences. There were demands for a centralized life science division both inside and outside of NASA in the late sixties. In 1970, NASA had to address the widespread dissatisfaction and inefficiencies of its adjunct life science task force by programmatic restructuring. John Pitts saw the reorganization of NASA's biomedical program in 1970 as a reflection of "the value of centralized coordination of life science programs." (Pitts 1985, p. 175)

The operational merits of centralized statecraft were presciently identified in seventeenth-century England by Thomas Hobbes, who maintained that "centralized definitions of reality are more effective and reliable as means of settling social disputes than decentralized definitions of reality which are based on collective witnessing." (Ezrahi 1990, p. 93). Boyle's experimental philosophy would dominate the Anglo-American political tradition for the next three centuries while Hobbes' hortatory advice of the strength of having a strong sovereign to avert *bellum omnium contra omnes* (the war of all against all) was buried in his treatise *Leviathan*. Recently, however, there is revived interest in assessing the relevance of Hobbesian moral and political philosophy to contemporary political affairs. Hobbes's disagreement with Boyle three centuries ago continues to matter given the ongoing difficulties of guaranteeing public order and securing global peace in the twenty-first century (Armitage 2006; Foisneau and Sorell 2004). Hobbs' astute observation of the advantage of having an absolute sovereign to arbitrate disputes in the seventeenth century foreshadowed the need for having a centralized coordination of life science components at NASA in the 1960s.

The US data was a test case for our inquiry in China. The historical circumstances of US civilian-military struggles in the context of Anglo-American political philosophy were intended to sharpen the consideration of the role of biophysicists in the Chinese space program mediated by Chinese political culture.

4.4 Historical Context of Chinese Political Culture

The above historical juxtaposition of the space program and statecraft is intended to bring into perspective the interactions of biology-related experts in Chinese space operations. While the decentralized political structure in the US could weaken the capabilities of federal cooperation and national deliberation, the centralized state in China had the bureaucratic capacity to offer a vertically integrated system for managing space activities. But it is worth asking: Did the Chinese centralized system really offer examples of better practices for disciplinary integration?

Skeptics would probably point to the infamous Anti-Rightist campaign that began in 1957 as counter-evidence. The mainstream characterization of the Anti-Rightist campaign is that it was mainly a hysterical witch-hunt against the "rightists" or "reactionaries" whom the Communist government unreasonably and unfairly identified as "enemies of the people" and severely punished with labor reeducation, imprisonment, or exile (Wang 2010b). What triggered the Anti-Rightist campaign was the preceding "Double Hundreds" movements (The Hundred Flowers and the Hundred Schools Movements) in which Mao legitimized the view that science had no class character following the speech made by Lu Dingyi (陆定一), then director of the Central Propaganda Department (Hu 2009). Many saw Mao's rhetoric as a way to smoke out intellectuals with bourgeois ideologies. Mao's wariness against intellectuals intensified when some scientists publicly stated that "the Communist Party does not know science and it cannot lead scientific work," and that "the leadership of the Communist Party is not beneficial for scientific development." (Hu 2009, p. 37). Met with these explicit outcries of direct challenge to the ruling authority of the party rather than constructive suggestions of policy advice, Mao took actions to crackdown "those who had misunderstood the limits of the earlier invitation to debate" as the erstwhile diplomat Kissinger (2011, p. 182) aptly described.

Some attributed the cause of the Anti-Rightist campaign to Mao's distrust and resistance against those who were more knowledgeable than him (Hu 2009). Mao's personal caprices and insecurity notwithstanding, the extent of the impact of the Anti-Rightist campaign has been disputed. Wang (2010b) reckoned that around 300,000 people were branded as "rightists" whereas one source put that figure at 550,000. The scope of influence on CAS scientists was even more uncertain. It is true that some scientists were sent off to labor camps or criticized in one way or another, but there were also explicit guidelines separating the treatment of scientists—particularly those with significant achievement—from that of social scientists. For example, one policy document granted special prerogative to natural scientists during the Anti-Rightist struggle: "…the Anti-Rightist struggle in the scientific sector should not be carried out in the same fashion as in social sciences. Anyone with significant achievements or those who returned after the Geneva conference should be protected." (Yao 1989, p. 455). A few historians of rocket technology in China interpreted this policy as ensuring that a group of outstanding but 'problematic' scientific practitioners could continue to work on atomic bombs and missiles.

Merle Goldman has offered her views on various reasons behind the selective exemption of scientists from the spate of political attacks in twentieth-century China. The boundary between scientific and nonscientific intellectuals is one crucial factor. That science was perceived as less ideologically subversive was captured by the belief that "because scientists worked with slide rules and equations, their work was ostensibly less related to political issues than was that of writers and social scientists which, by its very nature, challenged political control." (Goldman 1981, p. 136). Besides the perception of science as politically neutral, differential attitudes towards scientists were associated with the pragmatic value science afforded to modernization and production. Even at the revolutionary height of the sixties, Goldman argued that Mao "wanted to shield [scientists] from the kind of violent attacks that hit nonscientific intellectuals." (Goldman 1981, p. 138). Although the line separating scientists from nonscientists became fuzzier as the Cultural

Revolution kicked into high gear, it is fair to say that scientists in general were not reviled on the same scale as nonscientific intellectuals.

Moreover, historical studies of individual institutes at CAS reveal that scientists from the specific institutes under examination suffered relatively less than the general body of CAS during the Anti-Rightist campaign. Both Wang (2010b) and Schmalzer (2008) contended that scientists at their respective institutes of investigation—namely the Institute of Physics and the Institute of Vertebrate Paleontology and Paleoanthropology—did not suffer as much compared to their colleagues from other institutes at CAS. If that is the case, it would seem that the negative impact of the Anti-Rightist campaign on the science sector should be evaluated on the specific level of individual research institutes rather than on the generic level of CAS.

How did the scientific and engineering crews recruited in "Two Bombs, One Star" fare in the ostensibly anti-intellectual, anti-scientific political atmosphere of 1958? Some Chinese historians of combat and military sciences have argued that various political movements in the twentieth century had relatively minor impact on the conduct of "Two Bombs, One Star" because of its political significance. Since "Two Bombs, One Star" was presented as a matter of national security, it was packaged primarily as a political mission rather than as a scientific instrument. As Li, Liu and Xie incisively argued, "the project 'Two Bombs, One Star' figured high on the priority list of key political leaders because it was mainly considered as a (scientific) tool for achieving political ends. The project was given privileged treatment in many ways. This is why the project can be pushed forward successfully in the midst of the unstable political environments." (Liu et al. 2005, p. 123). Or as Lewis and Xue summarily remarked, "China accommodated its technology to politics, rather than the other way around." (Lewis and Xue 1988, p. xviii)

This perspective was further corroborated by another veteran in the history of China's aerospace industry. Besides concurring with the previous points on national priority and determination to advance missile research, Zhang Jun (张钧) approached the relationship between the political tumult and the space program with an appraisal of the political trustworthiness of intellectuals in space science:

> From the latter half of the fifties to the antecedence of the "Cultural Revolution," the "leftist" ideological inclination led to discrimination against scientific practitioners, depreciation of knowledge, and distrust of intellectuals. These behaviors were seen as expression of a "strong standpoint of class conflict." Yet the circumstances were different in the aeronautic frontier...Party leaders and cadre members from various ranks at the aeronautics outpost share a conception over time that rockets and orbital satellites are inseparable from intellectuals (Zhang 1986, p. 482).

It is important to note that Zhang Jun was not a China watcher from the outside, but an informed insider himself. He was the deputy director of the Seventh Ministry of Machine Building (Seventh MMB) between 1964 and 1968. The Seventh MMB was created in 1964 as the chief institutional body to oversee all bureaus in space science and technology, including the Fifth Academy. The Seventh MMB was the central unit in charge of the research, design, testing, manufacture, and management of missiles and rockets (Zhang 1986). The Seventh MMB was renamed the Ministry of Aerospace Industry in 1982 and Zhang Jun was appointed as director. Familiar with the military-civilian alliances in space operations, Zhang acknowledged that the space sector was inevitably involved in an array of political struggles in the second half of the twentieth century; yet Zhang also highlighted the extraordinarily high level of respect and care cadre leaders displayed towards intellectuals in the space industry. It is true that the revolutionary fervor did spread to the space sector, but party officials were chiefly concerned about the pragmatic reliance upon scientists for the research and manufacture of rockets and satellites— for these instruments were directly linked to national interests. Premier Zhou Enlai articulated the imperative of using advanced science and technology to consolidate national defense as a state priority in 1956 (Zhou 1997, pp. 370–371). The discourse paved the way for his later effort to insulate rocket scientists and engineers in the Seventh MMB from the attacks of radical factions by maintaining that the testing and launching of rockets is a matter of national security and reputation, and that anybody who interferes with the process is a traitor (Li 2005a).

Besides the Anti-Rightist campaign, the year 1958 is overcast with another political cloud that loomed over large-scale engineering projects; the over-ambitious Great Leap Forward was the next anthropogenic storm awaiting to unleash its calamity. Between 1958 and 1960, revolutionary optimism and production overestimation took a heavy toll on Chinese population. The extent of causality was compounded by the subsequent famine and an untimely drought. Judith Banister suggested that more than thirty million people died during the Great Leap Forward (Banister 1987).

The policy failure swiftly turned into an economic impasse. What is known as the "three hard years" forced the decision-makers in Beijing to put many state-led modernizing projects on hold. Much controversy flared up around the space-related projects, which were perceived as serving no immediate needs for people's well-being.

Whether to "continue" (上马) or to "abort" (下马) Mission 581 was a much disputed matter at the Bei Dai He Conference of National Defense and Industry. As the principal architect and national advocate of 'Two Bombs, One Star,' Marshal Nie found himself besieged by an assembly of conference participants who were opposed to the continuation of the space program. Many of them felt that China simply could not afford such a lofty project with flimsy connection to the suffering on earth. The issue was brought to the Politburo and Central Military Commission meetings at Mount Lu in the summer of 1960 where it attracted the attention of Chairman Mao (Gong 2006). In the past, Mao had tauntingly called nuclear threats and atomic weapons "paper tigers" that were subordinate to what was fundamentally a "people's war." (Lewis and Xue 1988, p. 242; Freedman 1981, p. 276). Nonetheless, he understood the strategic implications of space research for advanced technology and national security. One does not need to be a space enthusiast to appreciate the fact that controlling higher airspace is a military asset for overseeing operations on the ground. Mao therefore endorsed project 'Two Bombs, One Star' in spite of the opposition from other leaders. As Mao proclaimed, "we must make up our minds to develop cutting-edge technology. We cannot cut back or abort." (Zhang 1986, p. 16). Mao's declaration is a manifesto of the national determination to pursue advanced science and technology regardless of the financial difficulties.

Some might call it an irresponsible policy that was achieved at the peril of public welfare. This is certainly one way to look at the picture; another perspective suggests that there are always some economic excuses or social opposition to hold back the development of basic science. Socialist China between 1958 and 1960 was arguably not the best time for launching space programs in view of the enormous domestic difficulties. But if space programs had to be postponed until every hungry person had been fed, it could not have been launched at all. From this standpoint, the political resolution to continue the space program gave it the necessary impetus to stay the course. The technocratic cliques in the party welcomed Mao's decision. His stubbornness translated into a national tenacity to unify the space coalition.

In short, the decision to continue with the space program reflects the economic constraints and the various competing priorities facing China's decision-making authorities by the late fifties. Although the Anti-Rightist campaign and post-Great-Leap-Forward material shortage beclouded the space program, the high priority bestowed upon the space program in general provided a strong political imperative for the actualization of the space initiatives.

4.5 Biophysics and Mission 581

The political tone was settled as the space mission unfolded in earnest. In the foregoing paragraphs, I have briefly noted that biophysicists took part in Mission 581. I have also explored the nuanced dynamics between the civilian and the military sectors. The complex civilian-military interactions offered the backdrop against which the role of biophysicists could be assessed more clearly. But what was the exact responsibility of biophysicists in Mission 581? And how did Mission 581 transform biophysics?

The principal task of Mission 581 was to undertake the research and production of artificial satellites and sounding rockets. Researchers, experts and engineers from five sectors were enrolled in the planning and execution of Mission 581. Besides CAS and the Fifth Academy, workers, practitioners, and volunteers were mobilized from industrial sectors, higher-education institutions, and local research organizations to facilitate the project (*Science Times* 1999). The tactic was to commission a "big corps" to manage an unimaginably monumental task. It was a strategy of mobilizing mass efforts to overcome extraordinary obstacles.

The reliance on collective strength and willpower was not just a legacy of the Great Leap Forward but also an emerging paradigm of science in socialist China. One of the principal proponents of this collective mode of operation was Marshal Nie, who endorsed the means of using centralized power to enable national cooperation (Zhang 1986). In this model, the ideology of self-reliance aligned with the principle of national cooperation to inform the general policy for science and technology in the PRC. For example, a similar big corps was set up in 1960 to create synthetic insulin under Mission 601 (Xiong and Wang 2005).[4] The resultant synthetic bovine insulin with high vitality was lauded as a socialist achievement to crack the code of life (*People's Daily* 1967a, b).

The "big corps" style of management symbolized the myriad of participating units in Mission 581. The Institute of Biophysics was part of the CAS team, but the participation and contribution of biophysicists should not be overstated, to obscure the collective nature of the mission. Among the CAS alliances in Mission 581 were the Institute of Mechanics, the Institute of Electronics, the Institute of Biophysics, the Institute of Geophysics, the Institute of Automation, and the Institute of Physics. Scientists and engineers from CAS rallied a primarily research, academic-oriented force for Mission 581. Within the CAS squad, the Institute of Biophysics was charged with the task of launching biological sounding rockets while the Institute of Geophysics under the leadership of Zhao Jiuzhang was charged with the construction of geophysical sounding rockets.

Under the chairmanship of Qian Xuesen and the vice-chairmanship of Zhao Jiuzhang and Wei Yiqing, three institutes for general rocketeering were founded in 1958, namely the 1001 Institute of Satellites and Overall Design, the 1002 Institute of Control System Design, and the 1003 Institute of Satellite Payload Design. Of these, the 1001 Institute was mainly responsible for the supervision of the sounding rockets, which were non-orbital, non-recoverable small rockets that were intended for probing into the upper atmosphere for research purposes. Therefore, ground control, tracking and recovery—specialties of the other two rocket institutes—could be kept to minimum. The 1001 Institute was the chief partner unit working closely with the Institute of Biophysics and the Institute of Geophysics in launching the biological and meteorological sounding rockets respectively (Harvey 1998).

The 1001 Institute was also known as the Shanghai Institute of Machine and Electricity Design (SIMED) jointly administered by the Shanghai municipal government and CAS. Engineers from SIMED provided the mechanical shell while scientists from CAS determined the ends to which these probing rockets would be put. The partnership between SIMED and CAS worked this way: SIMED supplied rocket engineers and material physicists to design the hardware backbone of the rockets while scientists from CAS were left to take care of the interior systems and information-retrieval functions of the sounding rockets.

The T-7 rocket is the major type of vehicle in China's sounding rocket series. The letter "T" in the appellation is short for *tan kong* (探空), which means that all rockets in this family belonged to the sounding rocket series. The T-1 and T-2 rockets were untested models reverse-engineered from the German V-2

---

[4]Xiong and Wang (2005) gave a negative portrait of the 'big corps' mode of administration in synthesizing bovine insulin. They argued that the successful synthesis of protein depended on delicate laboratory skills rather than the crude enthusiasm of volunteers. The 'super big corps' (特大兵团) was later broken down and only the trained and sophisticated laboratory scientists were allowed to stay. It was these biochemists and pharmacists that finally rescued the project. They concluded that Mao's strategy of relying on the masses was not suitable for every line of work in science and technology.

(*Vergeltungswaffe*-2) rockets; the T-3 and T-4 were tested with liquid fluorine and methanol as the propellant, and the T-5 was launched with liquid oxygen and ethanol as the propellant. All previous experiments of the T-family rockets were unsuccessful until the second launch of the T-7M rocket ("M" denotes "model") on 13 September 1960. The T-7A (S1) and (S2) rockets (探空7号甲生物I型, II型火箭) used the same T-7 M rocket (T-7模型火箭) prototype but modified it with an animal capsule inserted into the nose cone section of the rocket (Li 2005b).

The close cooperation between SIMED and the Institute of Geophysics revolved around the assemblage of a meteorological sounding rocket for the purpose of studying the geophysical phenomena of the upper atmosphere. SIMED handled the engines and the types of fuels for firing the rockets while the geophysicists figured out what kind of barometers should be used to measure the atmosphere's pressure, density, temperature, wind, speed, and direction.

As for the biological sounding rockets, SIMED basically recycled the same type of missiles and fuels that they had previously tested on the earlier sounding rockets for meteorological exploration. After several successful runs with the radar and sensor-carrying sounding rockets, the biophysicists joined the crew to experiment with the idea of carrying a payload of living creatures to the upper atmosphere. This is the origin of the biological sounding rocket in Mission 581, for which the objective was not weather forecasting but the physiological measurement of the effects of rocket flight and space travel on living organisms. Biophysicists were invited to join the team in order to expand the scope of sounding rockets beyond meteorological applications.

The participation of biophysicists was important because the T-7A rocket was the first time a Chinese rocket was designed to carry biological payloads into space. The ancestors of the T-7A, i.e. T-7 and T-7 M rockets, were only intended to carry meteorological equipment such as radars and sensors. Engineers from SIMED needed input from biophysicists to devise ways of putting living things in the T-7A rockets. Table 4.1 presents the type of vehicles and the corresponding participation of biophysicists in the launch history of T-7 rocket flights:

The first biological sounding rocket T-7A (S1) was successfully launched in July 1964. T-7A weighed 1145 kg and stood 10.32 m tall, carrying 40 kg of payload and reaching 115 km.

In the second biological sounding rocket T-7A (S2), two Chinese space dogs named Xiao Bao (小豹) and Shan Shan (珊珊), along with other biological specimens[5] were carried to the ionosphere. The launching and landing were successful.

This successful news attracted the attention of the central government officials, most notably then-vice premier Li Fuchun (李富春) and other members of the Central Special Committee such as Zhang Jingfu. After the successful launches of the T-7A (S1 and S2) rockets, then-president Guo Moruo, Zhang Jingfu, and member of the

---

[5]The biological specimens that were put into the T-7A (S2) sounding rocket included a box of albino rats and test tubes containing fruit flies, actinomycin, parenzyme, lysozyme, pepsin, penicillin, phycomycin, bacteriophage etc.

Table 4.1 A launch history of Chinese T-7 rocket flights (emphasis my own)

Launch date	Vehicle	Launch site	Status	Participation of biophysicists
2/19/1960	T-7 M	Laogang	Successful	No
7/1/1960	T-7	Guangde	Successful	No
9/12/1960	T-7	Guangde	Successful	No
12/1/1963	T-7A	Guangde	Successful	No
7/19/1964	**T-7A(S1)[b]**	**Guangde**	**Successful**	**Yes**
6/1/1965	**T-7A(S1)[b]**	**Guangde**	**Successful**	**Yes**
6/5/1965	**T-7A(S1)[b]**	**Guangde**	**Successful**	**Yes**
7/15/1966	**T-7A(S2)[a]**	**Guangde**	**Successful**	**Yes**
7/28/1966	**T-7A(S2)[a]**	**Guangde**	**Successful**	**Yes**
8/8/1968	T-7/GF-01A	Jiuquan	Unknown	No
8/20/1968	T-7/GF-01A	Jiuquan	Unknown	No

Source Burgess and Dubbs (2007), Appendix W; and IBP-CAS (2008), p. 66
[a]Rockets carrying dogs, mice, and biological specimens
[b]Rockets carrying mice and biological specimens

biological academic division Tong Dizhou made a visit to the Institute of Biophysics in the company of Bei Shizhang. In January 1966, Bei chaired the First National Forum on biological satellites, and a conference on aeronautical medical science organized by the Ministry of National Defense. A report titled "the biomedical plan for a manned spaceship" was submitted to the Central Planning Committee (IBP-CAS 2008). In short, mission 581 opened a new page for the launch of the biological rockets; it was the first time biology-based scientists were actively engaged in the national missile program in China. The pioneering study of animals in spaceflight laid the necessary groundwork for manned space flight in the years to come.

As biophysicists were assigned the biological space mission, two puzzling questions stood out. First, why was the Institute of Biophysics picked for overseeing biological rocket flights? The Institute of Biophysics was in no way the only biology-related institute under the CAS framework for exploring the biological dimensions of space travel. In fact, it had just been inaugurated in 1958. Considering the newness of the institute and its staff, why would Chinese mission planners delegate a task involving high-altitude biological testing to biophysicists? Secondly, how did the biophysicists align itself for a task that fell more appropriately into the specialties of aviation medicine and space biology rather than biophysics? In short, how did biophysics and biological rocketry find each other?

The answer is that biophysicists sought biological rocket designers out, not the other way around. The delegation of biological rocketeering to the Institute of Biophysics was not a top-down decision from the central authority but a result of an unsolicited proposal from Bei Shizhang. Bei submitted a formal proposal to the CAS board of planners in May 1958. In this policy document, Bei convinced his comrades of the capabilities of his institute to conduct high-altitude biomedical research (Wang 2010a, b, pp. 247–249). Since biophysics was still a disciplinary blank slate, it offered more room to make the necessary preparations for a mission that required not just considerable expertise in atmospheric physiology but also seamless coordination with other units, primarily SIMED, in the design of the cockpit layout and life-support system in the flying capsule. Bei felt that his institute was up for this challenge due to the very interdisciplinary nature of biophysics. The launching of biological rockets incorporated in one flight both the physical aspect of testing rocket propulsion on factors of acceleration and deceleration and the biological aspect of measuring atmospheric effects on vital functions. Whoever ended up with the job would need knowledge in space flight and animal experimentation to satisfactorily execute the mission.

It would have been difficult to predict that the very same Bei Shizhang whose doctoral expertise was in experimental cytology would be entrusted with a mission in high-risk space biology. Bei had no proper training in rockets and aviation medicine. He was equipped with the knowledge on how to deal with the smallest enigmas of life but not the big question of flying animals into space. Neither Bei nor the institute under his leadership was the best candidate for the biological sounding rockets. But neither was anyone else in the rest of the country. When Qian Xuesen returned to China in 1955, he saw the bleak reality that he had to do everything from scratch (Chang 1995). Post-war China was at ground zero when it came to missile production and aerospace engineering. There were no launching pads, no observation consoles, no testing sites for high-altitude tasks of any kind. There was neither an existing pool of "missile-conscious" scientists nor "rocket-conscious" biologists for mission planners to choose from.

Besides the lack of scientific expertise from the Academy, the equipment and manpower of Chinese military were nowhere near the caliber of that in the USSR or the US in the late fifties. The Air Force of the People's Liberation Army did not have a strong team of aviation surgeons and combat clinicians as China did not emerge as a post-war air super power. In the 1950s, many Chinese scientists and engineers with very little or no former training in rocket engineering signed up to be chief engineers and supervisors for managing "Two Bombs, One Star." For example, Wang Xiji (王希季) and Yang Nansheng (杨南生) reportedly knew little about rockets and satellites when they were appointed as director and deputy director of SIMED respectively. Wang recalled that everyone felt they were "crossing the river by feeling the stones." (*Science Times* 1999, p. 181). Laymen transformed into experts as they learned about aerodynamics and biomedical aviation in the field with hands-on experience. Given the full extent of technological backwardness of postbellum China, delegating the space mission to a new institute with no existing space specialists was perhaps an understandable compromise.

But why did Bei want to venture into the risky and esoteric field of space biology? What could biophysics possibly gain from sending animals into space? Bei's hope was to strategically drive the disciplinary growth of biophysics with the completion of an important state-led mission. He was inspired by the prevailing slogan of "using mission to drive discipline." The idea was to take core political missions as the organizing principle to coordinate all relevant disciplines and sub-disciplines in a systemic fashion. In the case of biophysics, Bei considered using the space mission as a driving force to unify the research and educational aspects of biophysics. He explicitly sought to create a biophysics study program at USTC by drawing upon the connection with the space-and-rocket mission. In a USTC archival document, it was printed:

> This specialty (biophysics) studies the instruments for measurement and detection, sensing devices and methods for ground-based remote measurement and control in high-altitude exploration launched by artificial satellites or rockets. The orbit route and speed of rockets and satellites, as well as radar technologies for tracing long-distance target positioning are also subjects of research. Since it is closely related to national defense and requires a larger pool of talents, creating this specialty is of paramount importance.[6]

As illustrated in Chap. 3, USTC was governed by the operational principle known as "mission-drive-discipline" by mobilizing the resources and expertise from CAS. The policy intention was to drive the scientific "disciplines" at USTC through the specific "missions" undertaken by the research institutes at CAS. In this organizational framework, biophysics was the only founding department at USTC with a specialization in biology. The rest of the USTC departments were specialized areas in the physical sciences, engineering, and electronics with no crossings into the biological sciences. This is because all of the departments at USTC were expected to provide strategic service to the CAS-run project, "Two Bombs, One Star." All of the thirteen disciplines were in direct relationship to the making of atomic bombs, hydrogen bombs, and artificial satellites. The defense-oriented philosophy of USTC was unambiguously articulated in an internal reference document:

> The purpose of establishing the University of Science and Technology of China is to accommodate the research mission in areas of national defense and advanced science by educating scientific practitioners in these classified areas for national defense. The configuration of departments and specialties basically revolves around the needs for the nuclear energy and rocket industries.[7]

In other words, biophysics as a biology-related discipline was included in USTC because biophysics was a core unit in the satellite project. Biophysics was also of strategic interest to the atomic team for assessing the biological effects of radiation. The connection with bombs and space missions enabled the biophysicists to align their interests with the broader interests of the CAS-USTC bureaucratic structure.

The strategy of using the space mission as the rationale for promoting the biophysics program at the USTC does not end here. Creating a biophysics department was the first step, but where would the biophysics students go upon graduation? Job allocation is as important an index as students' enrollment in raising the reputation of an educational program. If the biophysics graduates could

---

[6]"Adjustment Plan for Seven Departments and Specialized Groups (Classified)" 七个系专业, 专门组调整方案(机密), OUA, USTC, Box 1963-WS-Y-33.

[7]"A Configuration Plan for the Departments, Specialties, and Specialization of USTC (Classified)" 中国科学技术大学系, 专业、专门化设置计划表(绝密), in the Office of University Archive (OUA), USTC, Box 1960-WS-Y-21.

not apply what they have learned into practice, then the mission-drive-discipline policy was not as effective an attempt as those in decision-making positions had hoped. In elaborating the policy merits of "mission-drive-discipline," Zhang Jingfu, then deputy director of CAS, suggested:

> there were many positive aspects of driving disciplines with missions: 1) the research objective is clearly identified with a close correlation between theory and practice; 2) the details are concrete enough to draw mass support; 3) mission is like a red thread that connects all the beads of specialized knowledge together, with strong organizational vitality; 4) people with high and low operational levels could benefit from each other to enable public participation in science research activities (Zhang 2009, p. 55).

The fourth point in Zhang's elaboration translates into another aspect of the relationship between mission and disciplines. The policy shorthand "mission-drive-discipline" was to be matched by a complementary catchphrase: "discipline-facilitate-mission." One way to assess how "discipline-facilitate-mission" is to see whether the disciplinary trainees actually served in the mission. The satellite mission offers a policy imperative to create a biophysics department at USTC. But did the biophysics graduates provide adequate service to the satellite mission?

Besides the researchers at the biophysics institute, biophysics graduates from USTC were also recruited to work in the space mission. Most biophysics students (under the "cosmobiology" major) from USTC were directly allocated to the Institute of Biophysics upon graduation. In a special anthology recently published by IBP-CAS, several of the USTC alumni recalled their work experiences in the space mission between 1963 and 1966.[8] In brief, the biophysics team in Mission 581 was comprised of both veteran biophysicists and fresh graduates from not just USTC but other higher-educational institutions as well (IBP-CAS 2008, pp. 229–234).

Moreover, the teachers responsible for teaching some of the cosmobiology courses at USTC were directly appointed from the cosmobiology research unit at the Institute of Biophysics. The cosmobiology curriculum was designed to expose students to basic ideas that were necessary for putting together the biological sounding rockets. Deng Jiaqi (邓家齐) and Li Zhenxiang (李祯祥) were the core teaching staff for the cosmobiology major at USTC. Deng taught three courses, namely "Cosmobiology in closed environments," "The problems of radiation in perigee altitude," and "Explosive Cabin Decompression" while Li was responsible for two courses, namely, "Biomechanics in space" and "Uranobiology" or "Exobiology."(IBP-CAS 2008, pp. 216–219). In short, the teaching content and job placement of biophysics at USTC were in line with the practical needs of the satellite mission.[9]

---

[8]The seven USTC biophysics graduates who were featured in the special volume were Jia Kepu (贾克朴) and Pei Jingchen (裴静琛), Yang Tiande (杨天德), Ma Zhijia (马治家), Xue Yueying (薛月英), Chen Mei (陈湄), and Teng Yuying (滕育英).

[9]For details on the biophysics teaching program at USTC, see Luk (2015).

4.6 Concluding Remarks

The answer to the first driving question underpinning this chapter—the relevance and contribution of biophysicists to the Chinese space program—is that the biophysicists were part of Mission 581. Biophysicists were chosen to shoulder the responsibility of launching the biological sounding rockets in view of the crude conditions in mid-twentieth century China. The successful execution of Mission 581 depended on the intimate cooperation between scientists dispatched from CAS (of which biophysicists were simply one sub-group) and military officers appointed by the Second MMB and later the Seventh MMB. Some of the dialectical exchanges between CAS and the Second MMB exposed the fact that the civilian-military components of Mission 581 were neither perfectly dovetailed nor completely misaligned. It is useful to consider the bigger historical and political contexts in which the civilian-military alliance was situated in China as opposed to the US. The transformation of the civilian-military relationship was embedded in each country's specific historical matrix and bureaucratic orthodoxy. In particular, the intertwining of military services and civilian agencies illustrated the intricate relationships between science and the states. The rivalry between NASA and the Air Force in managing its biomedical capabilities is not a coincidence but a reflection of the normative political structure in the US.

As for the second question concerning the contribution of the space mission to biophysics, it is evident that the close collaboration between the defense ministry and the Academy provided the financial support and political legitimation for this new discipline to flourish. Between 1958 and 1966, the biophysicists at CAS and USTC were directly involved in the national military programs in rockets and missiles. The peak of its engagement with the techno-nationalistic program in Maoist China was its central role in the launch of biological rockets, T-7A (S1 and S2) between 1963 and 1966. The mission transformed the research and teaching of biophysics through the policy of "mission drives discipline, discipline facilitates mission." The intersection between space missions and the biophysics discipline offered a case in point to interrogate the relationship between sciences and the state in the PRC. In Maoist China, the relationship between science and the state that evolved as a result of the missile-cum-space program proved to be a determining factor in the growth of new or otherwise marginal sciences. Biophysics in particular profited from the opportunities for expansion and innovation provided by the state-led rocket program.

References

Armitage D (2006) Hobbes and the foundations of modern international thought. In: Brett A et al (eds) Rethinking the foundations of modern political thought. Cambridge University Press, Cambridge, pp 219–235

Banister J (1987) China's changing population. Stanford University Press, California

Burgess C, Dubbs C (2007) Animals in space: from research rockets to the space shuttle. Springer, Praxis

Chang I (1995) Thread of the silkworm. Basic Books, New York

Corliss W (1971) NASA sounding rockets, 1958–1968: a historical summary. NASA, Washington

Ezrahi Y (1990) The descent of Icarus: science and the transformation of contemporary democracy. Harvard University Press, Cambridge

Foisneau L, Sorell T (eds) (2004) Leviathan after 350 years. Oxford University Press, Oxford

Freedman L (1981) The evolution of nuclear strategy. St. Martin's Press, New York

Goldman M (1981) China's intellectuals: advise and dissent. Harvard University Press, Cambridge

Gong XH (巩小华) (2006) Inside the decision-making world of Chinese space industry (中国航天决策内幕). Chinese Literature and History Press, Beijing

Graham L (1974) Science and philosophy in the Soviet Union. Vintage Books, New York

Harvey B (1998) The Chinese space programme: from conception to future capabilities. Wiley-Praxis, Chichester

Hu HK (胡化凯) (ed) (2009) A selected collection of documents on animadvert on science in China during 1950–1970s, vol 1 (20世纪50-70年代中国科学批判资料选, 上卷). Shandong Education Press, Jinan

IBP-CAS (2008) Flying dogs in the sky: a documentation of Chinese biological experimental rockets. Science Press, Beijing

Kissinger H (2011) On China. Penguin Books, New York

Lewis J, Xue LT (1988) China builds the bomb. Stanford University Press, California

Li CZ (李成智) (ed) (2005a) A draft history of the development of space technology in China, vol 1–3 (中国航天技术发展史稿-上中下册). Shandong Education Press, Jinan

Li DY (李大耀) (2005b) Cherish the memory of the 45th anniversary of the first successful launching a series of the T-7 sounding rocket (为中国开发太空探路, 创中国首飞太空记录—纪念T-7系列探空火箭首次发射成功45周年). Spacecr Recovery Remote Sens 航天返回与遥感 26(3):1–4

Link MM (1965) Chapter 4, NASA long-range life sciences program: the kety committee. In: Space medicine in project mercury, NASA, DC. http://history.nasa.gov/SP-4003/ch4-4.htm. Accessed 21 Dec 2014

Liu JF (刘戟锋), Liu YQ (刘艳琼), Xie HY (谢海燕) (2005) The project of 'Two Bombs, One Satellite:' a model of the big science (两弹一星工程与大科学). Shandong Education Press, Jinan

Luk CYL (2015) Building biophysics in mid-century China: the university of science and technology of China. J Hist Bio 48(2):201–235

Mackowski MP (2002) Human factors: aerospace medicine and the origins of manned space flight in the U.S. Ph.D. dissertation, Arizona State University

People's Daily (1967a) China achieves world's first total synthesis of crystalline insulin. People's Daily, 1 January

People's Daily (1967b) Using Mao tse-tung's thought to open the gate to 'The Enigma of Life.' People's Daily, 1 January

Pitts J (1985) The human factor: biomedicine in the manned space program to 1980. NASA, Washington

Rasmussen N (1997a) Picture control: the electron microscope and the transformation of biology in America, 1940–1960. Stanford University Press, California

Rasmussen N (1997b) The mid-century biophysics bubble: hiroshima and the biological revolution in America, revisited. Hist Sci 35:245–293

Schmalzer S (2008) People's Peking Man: popular science and human identity in twentieth-century China. University of Chicago Press, Chicago

Science for the People (1973) China: science walks on two legs. Discus Books, New York

Science Times (科学时报) (ed) (1999) May history remember them: Chinese scientists and 'Two Bombs, One Star' (请历史记住他们: 中国科学家与'两弹一星'). Jinan University Press, Guangzhou

Song J (宋健) (ed) (2001) Biographies of pioneers of 'Two Bombs, One Star', vol 1 and 2 ('两弹一星'元勋传-上下卷). Tsinghua University Press, Beijing

Wang ZY (2008) In sputnik's shadow: the president's science advisory committee and cold war America. Rutgers University Press, New Jersey

Wang GY (2010a) Bei Shizhang: a biography. Science Press, Beijing

Wang ZY (2010b) Physics in China in the context of the cold war, 1949–1976. In: Trischler H, Walker M (eds) Physics and politics: research and research support in twentieth century Germany in international perspectives. Franz Steiner Verlag, Stuttgart, pp 251–276

Xiong WM (熊卫民), Wang KD (王克迪) (2005) Synthesize a protein: the story of total synthesis of crystalline insulin project in China (合成一个蛋白质:结晶牛胰岛素的人工全合成). Shangdong Education Press, Jinan

Yao S (1989) Chinese intellectuals and science: a history of the Chinese academy of sciences (CAS). Sci Context 3(4):455

Zhang J (张钧)(ed) (1986) Aerospace industry in contemporary China (当代中国的航天事业). Chinese Social Science Press, Beijing

Zhang ZH (张志辉) (2009) Selected literature on China's great leap forward in science and technology, vol 1 (科技'大跃进'资料选–上). Shandong Education Press, Jinan

Zhou EL (1997) Selected military writings of Zhou Enlai (周恩来军事文选). Military Academy of Sciences of People's Liberation Army of China Press, Beijing

Chapter 5
Conclusion: A Discipline Defined by Doing

Abstract This chapter takes a bird's eye view of the disciplinary landscape of biophysics in contemporary China by summarizing the major characteristics and the unique trajectory it underwent in twentieth-century China. Biophysics is a field of science that is broadly conceived with an amorphous definition, i.e. "defined by doing." Because of its inherent flexibility, biophysics is an ideal candidate to examine how supra-scientific considerations shape the varying content and context of this scientific discipline. I argue that biophysics offered a good vantage point to analyze how factors in different historical and cultural circumstances enabled the formation of science. The preceding chapters showed that biophysics gained traction in the PRC because its leaders devised strategies to gradually integrate the lives of ordinary people with the political and social developments of its time. It is the intersections between biophysics and the state, science and politics, disciplines and missions, cell biology and physical intervention that make biophysics a useful analytical lens to assess the history of science and technology in contemporary China.

Keywords China's scientific community · Disciplinary history of biophysics · Disciplines and missions · History of science and technology · Science and the state

Since this study puts a narrow focus on biophysics, the reader might incorrectly infer that biophysics was the hub around which all of contemporary China's interests in science and technology revolved; this was simply not the case. The disciplinary history of biophysics was only part of a larger and more complex phenomenon of science's interaction with the state. The somewhat privileged position of biophysicists in building their disciplinary structure and community was only part of the politics surrounding the governance and autonomy of science.

Although biophysics was not a policy priority for the key party leaders, it was an area in which several fundamental issues intersected. At the center of the story is a field of science that is broadly conceived with an amorphous definition, i.e. "defined by doing" as one biophysicist put it. Because of its inherent flexibility, biophysics is an ideal candidate to examine how supra-scientific considerations shape the varying

© The Author(s) 2015
C.Y.L. Luk, *A History of Biophysics in Contemporary China*,
SpringerBriefs in History of Science and Technology,
DOI 10.1007/978-3-319-18093-9_5

content and context of this scientific discipline. I argue that biophysics offered a good vantage point to analyze how factors in different historical and cultural circumstances enabled the formation of science. The preceding chapters showed that biophysics gained traction in the PRC because its leaders devised strategies to gradually integrate the lives of ordinary people with the political and social developments of its time. It is the intersections between biophysics and the state, science and politics, discipline and mission, cell biology and physical intervention that make biophysics a useful analytical lens to assess the history of science and technology in contemporary China.

Throughout the twentieth century, Chinese biophysicists gradually established a scientific discipline. Between 1958 and 1964, biophysicists created a department of biophysics and transformed the biophysics study program; between 1966 and 1971, they went from launching the biological sounding rockets to being a part of the military-industrial complex; between 1972 and 1980, they went from leading the US—China scientific exchange to launching a specialized journal and a professional society. What the history of biophysics in contemporary China demonstrates is the concrete steps Chinese biophysicists took to realize their dream of building a scientific discipline.

Large-scale institutionalization of science and technology in modern China became more apparent after 1949. The policy imperative and nationalist sentiment to pursue wealth and power via the avenue of science and technology offered a favorable environment for new scientific fields to develop. The disciplinary history of biophysics was deeply embedded in this unique national, social, and historical context of modern China.

During the Cold War, biophysicists accommodated their discipline to the needs and expectations of the state. This is clearly illustrated in the priority given to Mission 581—a mission with specialized knowledge unfamiliar to Bei or any existing biophysicists at the time. But they were willing to create new research facilities and training programs as needed in order to fulfill the mission's requirements. Participation in "Two Bombs, One Star" provided biophysicists with the aura of military connections and the prestige of laying the groundwork for subsequent human spaceflight.

Accompanying biophysicists' commitment to the space-and-rocket mission is their ambition to create an independent biophysics educational program at the newly established USTC. The concentration of the biophysics program at USTC reflected biophysicists' roles in the space mission. Between 1963 and 1966, those who majored in cosmobiology at USTC were recruited into the sounding rocket project and worked alongside rocket engineers, nuclear scientists, and military servicemen.

The Mao-era episodes of biophysicists working with and negotiating with the Ministry of National Defense illustrated how scientists lived with military officers and learned to adapt military priorities to their disciplinary interests, especially during the Cultural Revolution. Mission did not just "drive" disciplines by giving them the opportunity to thrive at USTC; it also protected mission-oriented disciplines from political critiques when such needs arose. The completion of "Two

Bombs, One Star" in the post-Mao period precipitated a movement toward professionalization among biophysicists. Professional interests replaced national interests after the completion of the space-and-rocket mission. As the partnership between state-delegated mission and scientific disciplines faded, the slogan "mission drives disciplines, disciplines facilitate mission" fell out of fashion. Bei Shizhang and other leading biophysicists sought new ways to sustain their discipline, and ultimately they were able to seize the incentives offered by the changing political circumstances in the late 1970s to enhance the professional status of biophysics.

Under Deng, the scientific community was allowed to pursue the internal developments of their disciplines, and while self-governance and/or self-determination were not promised, Deng-era scientists did enjoy an increased access to professional knowledge that was previously seen as "elitist" and "bourgeois." Readers of *PIBB* were exposed to accounts of professional development of biophysics around the world and this in turn prompted the domestic professionalization of biophysics, which had begun in the post-Mao period with the establishment of the Biophysical Society of China in 1980. Throughout the 1980s, the Biophysical Society of China was actively pursuing opportunities for academic exchange with and recognition by the global scientific community. It became a member of the International Union of Pure and Applied Biophysics by 1984 and the host for its annual meeting in 2011.

In order to achieve a more balanced assessment of the disciplinary status of biophysics in contemporary China, it is necessary to identify both its strengths and weaknesses. Chapters 3 and 4 have examined the institutional structure and the process of developing a unified research-cum-teaching policy driven by the space mission. One thing that has become clear from these chapters is that the major factor shaping the disciplinary formation of biophysics in contemporary China appears to be the political authority of its leading scientists, most notably that of Bei Shizhang. However, it is important to bear in mind that Bei's reputation is not without blemish. The analysis of Bei's scholarship and academic background in Chap. 2 illustrated the lack of academic acceptance of his scientific proposition. His lack of scholarly achievement can be attributed to his neo-Lamarckian theoretical conviction, among other plausible factors.

Another important qualification for this study is the distinction between institutional patterns and research outcomes. The institutional structure of biophysics does not necessarily have a linear relationship with the effectiveness of its research enterprise. Just because biophysicists are concentrated in one place under the supervision of centralized bureaucracy does not guarantee the output of high-quality biophysical research. As a matter of fact, the one original research pursued by Bei Shizhang under the rubric of "cell reformation" was not well received by the international scientific community. As it stands now, the history of biophysics in China seems to be largely a service discipline for the space program without significant contribution to the production of scientific knowledge.

In the past, the leading Chinese biophysicists might not have been world-renowned scientists, but they were good at drawing existing resources in the political arenas to facilitate the growth of a marginal discipline. As a visionary institutional builder, Bei's most important legacy was the discipline he founded. His primary achievement may well be seen as having constructed a disciplinary foundation in which a relatively broad range of interests are embedded.

In the final analysis, although the historical trajectory biophysics underwent in contemporary China does not endow its practitioners with an absolute advantage in pursuing basic research, I suggest that the disciplinary landscape of biophysics can be considered as "optimal" not because it is the "best" for conducting research, but rather in the sense that it is characterized by a process that can produce policy coordination and take a decent range of inputs into account. Whether this "optimal" feature can be transformed into research proficiency depends on the capacity of the biophysicists to mobilize the built-in strengths to overcome its apparent weaknesses.

Index

© The Author(s) 2015
C.Y.L. Luk, *A History of Biophysics in Contemporary China*,
SpringerBriefs in History of Science and Technology,
DOI 10.1007/978-3-319-18093-9